POLLUTION CONTROL AT ELECTRIC POWER STATIONS

Comparisons for U.S. and Europe

By

Richard D. Brown
Robert P. Ouellette
Paul N. Cheremisinoff

230 Collingwood, P.O. Box 1425, Ann Arbor, Michigan 48106

Library of Congress Catalog Card Number 82-45987
ISBN 0-250-40618-7

Manufactured in the United States of America

Butterworths, Ltd., Borough Green,
Sevenoaks, Kent TN15 8PH, England

PREFACE

The purpose of this book is to summarize pollution control expenditures and environmental regulations and standards pertaining to electric power stations in the Federal Republic of Germany, France, the United Kingdom and the United States. As far as possible, the pollution control expenditures are broken down into capital investment costs and operation and maintenance costs. Additional division into pollution control and power source categories is made where data are available. Information on pollution control research and development costs also is provided. Where possible, discussion focuses on the period 1970–1980.

It is difficult to relate capital costs of compliance to pollution control regulations. For example, in complying with the New Source Performance Standards of the U.S. Clean Air Act, utilities can use low-sulfur coal, or high-sulfur coal and scrubbers. The first solution imposes substantial operating costs, but limited capital costs. The second solution imposes both capital and operating costs.

It is often difficult to determine how much of a particular capital investment is being made for environmental reasons and how much would be made in the absence of any environmental standard. For example, some power plants have installed cooling towers to meet specific state water quality guidelines or in anticipation of federal thermal guidelines, but others installed similar equipment for engineering or economic reasons. When surveys are conducted to ascertain why utilities are building cooling towers, the answer often is because of both environmental and economic reasons, making it very difficult to determine how much of the investment was driven by compliance with regulatory programs and how much would have occurred in the absence of these programs.

Utility companies in general have not been prepared to account for the total costs of regulation. Although they can identify and isolate many administrative and operational costs of regulation, the effects on consumer rates are nearly impossible to assess accurately.

Regarding foreign data on pollution control expenditures, certain constraints exist. These relate to incomplete cost breakdown by fuel type; inconsistencies in itemizing research and development, environmental assessment, and control technology costs; incomplete data for the entire 1971–1980 period; data not available from various countries for particular categories; and data available only in aggregate for all utilities.

The reader should note the datedness of certain information as indicated by the citations. Environmental regulations are almost always in a state of flux. New regulations may have been promulgated in the Federal Republic of Germany, France and the United Kingdom that supersede those cited in the text.

Care was taken to indicate the date on which data were accumulated or the report date for data obtained from reports by the Organization for Economic Cooperation and Development (OECD).

We appreciate help from representatives of Electricité de France, Umweltbundesamt (Federal Republic of Germany), and the Central Electricity Generating Board (England) for providing needed information and acknowledge the sponsorship of the Overseas Electrical Industry Survey Institute, which made this book possible.

Richard D. Brown
Robert P. Ouellette
Paul N. Cheremisinoff

Brown

Ouellette

Cheremisinoff

Richard D. Brown is on the department staff at The MITRE Corporation, and specializes in the development and management of research programs relating to the health and environmental effects of energy technologies. He holds a BS in Zoology from Indiana University and a PhD in Limnology/Water Pollution from the University of Delaware. Before his association with The MITRE Corporation, he established the Department of Water Sciences at Jack McCormick and Associates, Inc., an environmental consulting firm. He also established graduate-level courses in advanced limnology and water pollution as a member of the graduate faculty at Indiana State University.

During 1979–1981 Dr. Brown served as Executive Secretary for the Federal Interagency Committee on the Health and Environmental Effects of Energy Technologies. He has presented testimony before Congress, at hearings at the federal, state and local levels, and at court proceedings. He is author of more than 75 publications.

Dr. Brown provided assistance to the Government of Malaysia in the development of a national environmental quality program. He has participated in a number of international symposia. He has recently completed a comparison of national environmental policies and research programs for the Ministére de l'Environnement, France. For the Japanese Overseas Electrical Industry Survey Institute, he has studied national pollution control strategies and costs relating to the production of electricity.

Robert P. Ouellette is Technical Director of the Environment Division of The MITRE Corporation. Dr. Ouellette has been associated with MITRE in varying capacities since 1969. He is a graduate of University of Montreal and received his PhD from the University of Ottawa. A member of the American Statistical Association, Biometrics Society, Atomic Industrial Forum and the National Science Foundation Technical Advisory Panel on Hazardous Substances, Dr. Ouellette has published numerous technical and scientific papers and books on a wide variety of subjects. He is co-author and co-editor of the comprehensive Electrotechnology survey series, as well as other books published by Ann Arbor Science.

Paul N. Cheremisinoff is Associate Professor at the New Jersey Institute of Technology, Newark. A registered Professional Engineer and consulting engineer, he has been a consultant on environment/energy/resources projects for The MITRE Corporation. An internationally known scholar and researcher, he is author/editor of many Ann Arbor Science publications in engineering/energy/environmental control, including *Pollution Engineering Practice Handbook*, *Carbon Adsorption Handbook* and *Environmental Impact Data Book*.

CONTENTS

TABLES

CHAPTER 1

OVERVIEW

POLLUTION CONTROL EXPENDITURES

Of the countries addressed in this book, the electric utility industry of the United States commits the highest percentage of capital investments in pollution control equipment, relative to the total capital costs of power stations. The overall percentage of total capital costs committed to pollution control by the U.S. electric utility industry is about 10%. The percentage for the United Kingdom is 5–10%. France and the Federal Republic of Germany commit about 2.6 and 1.8%, respectively (Table 1-1). Electric utilities in the Federal Republic of Germany and the United States appear to commit about 45% of the total pollution control budget to air pollution abatement. In France, most pollution control expenditures are for water pollution abatement. This is largely due to the relatively high use of nuclear power, where most air pollution controls are not needed and the amount of heated water discharge is high compared to fossil-fueled power plants. Pollution control data are not readily available for the United Kingdom, but most expenditures appear to be related to water pollution control and improvement of esthetic amenities (see Appendix A).

Sulfur dioxide control policies have been implemented recently in three of the four countries addressed in this book. The estimated annual control costs committed for desulfurization of flue gas or high-sulfur (average 3.5% sulfur) residual fuel oil average (1980 U.S. $) are $409/ton of sulfur dioxide removed in the Federal Republic of Germany and $343/ton removed in the United Kingdom [OECD 1981]. France does not have a commitment to desulfurization, because of a program to switch dependence from fossil fuels to nuclear energy as a power source [OECD 1981].

Table 1-1. Percent of Capital Investment in Pollution Control Equipment Relative to Total Capital Costs of Power Stations

	FRG, 1975[a]	France, 1980[b]	United Kingdom, 1974	United States, 1980[c]
Air	0.82	0.38	0.39[d]	4.52
Water	0.46	1.06	2.60[e]	1.89
Solid Waste	0.14	NR[f]	NR	0.62
Noise	0.37	0.41	0.0 [g]	0.09
Esthetics	NR	0.04	NR	2.31
Miscellaneous	NR	0.74[h]	NR	0.87[i]
Overall Percent of Total Capital Investment	1.8	2.6	5–10[j]	10.3

[a] Aggregate data for all utilities, including electricity, gas, water supply and district heating.

[b] Power station pollution control costs as a percentage of total utility investment (including transmission costs).

[c] Based on actual in-service capital investments, rather than on costs of construction in progress or new investments for a given year.

[d] Capital cost of modern coal-fired plant attributable to tall chimney and 99.3% efficient precipitator is 3%. The addition of sulfur dioxide control equipment could increase this value to as much as 3.9 or an average of 0.39/year over a 10-yr payoff period.

[e] This value could be as much as 6.7 times the cost of investments to control air pollution based on long-term (1971–1980) ratio of air/water investment costs for all industries [OECD 1974].

[f] NR = data not reported.

[g] Noise control represents 5% of the cost of new gas turbine stations. These represent less than 1% of total generating capacity.

[h] Radiation control.

[i] Additional plant capacity and miscellaneous environmental protection facilities.

[j] Total cost of pollution control as a percentage of plant capital costs [Barrett 1982].

Costs in the United States for the same period are estimated to be $212–399/ton of sulfur dioxide removed, depending on control technology used [EPA, unpublished data].

Costs (1980 U.S. $) among the countries studied seem to have stabilized to $300–400/ton of sulfur removed by desulfurization (primarily from flue gas). In the early stages of the development of these processes, there were considerable differences in the estimated capital and operating costs among countries (Table 1-2) [OECD 1973].

Overall expenditures and legislation for pollution control in the United Kingdom appear to follow the U.S. lead in terms of pollution abatement activities. Legislation similar to the U.S. Clean Air and Water Acts is being implemented, and pollution control expenditures are rising accordingly in the European countries. In the United States, the major increases

Table 1-2. Desulfurization Costs (1973 U.S. $/ton of sulfur removed) [OECD 1973]

Country	Capital Costs	Operating Costs
Federal Republic of Germany		310[a]
France	NA[b]	NA
United Kingdom	500	180
United States	2500	480

[a] The operating costs for Germany include capital costs. These are based on a 4000-hr/yr load factor at full capacity. Investment costs are approximately 30% of operating costs.
[b] NA = not applicable. France has had no national program directed toward desulfurization, because of the development of nuclear power as an energy source.

in capital expenditures in response to environmental regulations will be in the 1980s (Table 1-3).

AIR POLLUTION ABATEMENT TECHNIQUES

The countries addressed in this book have applied various strategies to control air pollution from power plants. The strategies include fuel substitution, desulfurization of fuel oil, flue gas scrubbing and desulfurization, physical coal cleaning, particulate collection equipment (mostly electrostatic precipitators), combustion control, tall stacks, gasification of coal, gas turbines, fluidized bed combustion, nuclear energy, conversion from "yellow flame" to "blue flame" oil burners, and conservation of energy. The following discussion highlights important aspects of some of these techniques as they relate to the countries addressed in this book.

Fuel Substitution

Fuel substitution is a convenient method for reducing pollution. Gas and distillate oil, although relatively costly, are ideal substitutes. Solid fuel yields larger particles and more nitrogen oxides than does residual fuel.

European coal is preferred over residual oil as it releases less sulfur dioxide when burned. This relationship is the reverse in the United States, where most coal has a higher sulfur content. In the United States, high-sulfur coal is replaced by low-sulfur coal, when availability and price are favorable. Some European countries have followed a guideline of replacing coal by 1% sulfur oil in power stations [OECD 1973].

Table 1-3. Summary of Impacts of Environmental Regulations on Capital Expenditures, 1970–1990 (10^9 $, 1980) [ICF 1980]

Regulations	Increase in Capital Expenditures (1970–1978)[a]	Increase in Capital Expenditures (1979–1990)[b]	Total Increase in Capital Expenditures (1970–1990)
Clean Air Act Amendments of 1970	3.1	25.5	28.6
Clean Air Act Amendments of 1977		22.8	22.8
Federal Water Pollution Control Act of 1972 and Surface Mining Control and Reclamation Act of 1977	6.3	6.8	13.1
Total Impacts of Current Environmental Regulations	9.4	55.1	64.5

[a] Approximately $0.3 billion of these expenditures were projected to occur in the 1965–1969 period.

[b] Capital expenditures between 1979 and 1990 in the absence of environmental regulations are projected to be about $508 billion.

Desulfurization

Desulfurization of fuel oil is an option chosen by most European countries, where residual fuel is desulfurized to 1% sulfur. Desulfurization also takes place in the Caribbean, to supply low-sulfur fuel oil to the U.S. East Coast.

In the United States, flue gas desulfurization (FGD) by wet scrubbing is being used. However, because of the large amounts of liquid sludge that need disposal, regenerable or recycled control types of FGD are expected to be implemented in the 1980s.

Physical Cleaning of Coal

Coal cleaning appears to have been used widely only in the United States. However, the Federal Republic of Germany has been steadily increasing the use of this practice since 1974. Of the nearly 500 million tons of coal used by the U.S. utility industry in 1975, 20% was cleaned. From four carloads of raw coal, about one carload of impurities can be removed by physical coal cleaning. Desulfurization can be accomplished during cleaning by gravity separation, flotation and magnetic separation to remove pyrite sulfur. Organic sulfur can be removed by chemical coal cleaning, but the current cost of various processes under development makes its removal prohibitive [Brown 1979].

Tall Stacks

This technique has been used in France and the United Kingdom to reduce ground-level sulfur dioxide concentrations by dispersion into the upper atmosphere. To take full advantage of this technique, it is used with fuels with the highest possible sulfur content, to allow lower-sulfur fuels to be burned in locations where tall stacks cannot be used.

The practice followed in the United Kingdom since 1958 for large, conventionally fueled power stations has been to discharge flue gases through a single stack exceeding 150 m in height at an effective velocity of 15–23 m/sec and a temperature of 110–150°C. Under these conditions, the momentum and buoyancy of flue gases carry them above any temperature inversions of the atmosphere. Measurements have shown that British power stations do not contribute appreciably to ground-level sulfur dioxide concentrations, even when meteorological conditions are such that pollutants discharged at lower levels tend to accumulate. In France, the exit velocity is somewhat higher (20–30 m/sec), to ensure that ground-level sulfur dioxide concentrations do not exceed 0.25 mg/m^3 [OECD 1973].

Disposal of sulfur dioxide into the atmosphere by high stacks is a very controversial subject. It is claimed that it can be transported over large distances and can come down in the form of acid rain, thus damaging buildings and vegetation, and causing increases in the acidity of lakes and groundwater. The United States explicitly rejects tall stacks as a satisfactory solution to international pollution problems [OECD 1973].

DIFFERENCES IN POLLUTION CONTROL STRATEGIES

The United States, with its relatively large land area and long coastline, allows power stations to be sited in remote locations. Most (about 80%) electric utilities are in private ownership. Varying state and federal pollution control regulations have been effective in abating pollution from power plants for more than a decade. Overseeing of environmental protection is provided by the federal government.

The United Kingdom, densely populated, with a long history of land-use planning, has a nationalized electrical supply system. Pollution control authority is a mixture of centralized and decentralized control, the latter administered by regional and local government officials, collectively termed the Inspectorate.

The Federal Republic of Germany has a relatively high population density and a relative lack of cooling water. Its short coastline, with the water depth increasing slowly, and with few rivers suitable for direct cooling, has meant that cooling towers are required for nearly all large coal-fired and nuclear power plants. Pollution control authority is shared by federal and state agencies.

The French, in many respects like the British, have a long coastline and a decentralized pollution control program. In France, the program is largely administrated by local and regional prefects.

In the Federal Republic of Germany and the United Kingdom, stringent urban air quality standards for sulfur dioxide have led to an improvement in ground-level air quality in polluted areas. This improvement has been accomplished by the use of tall stacks in urban areas and the siting of plants in rural areas. These actions have increased air pollution elsewhere, created international transboundary movement of pollutants, and created public opposition to the increased impact on rural esthetic amenities.

U.S. pollution control legislation places much emphasis on ambient standards. The emphasis of British and German legislation is on emission standards. The Federal Republic of Germany has an additional set of standards, termed "immission" standards, that relate to air quality within the immediate vicinity of a stack. France currently emphasizes integrated pollution control through planning and siting of new facilities.

TRENDS IN THE USE OF FUELS FOR POWER PLANTS

Coal

Almost all new nonnuclear conventional power plants in the United States are expected to be fueled by coal. New coal-fired electricity-generating capacity in Europe is expected to be added mainly in the United Kingdom and the Federal Republic of Germany. France is expected to develop nuclear power instead of relying on fossil fuels [OECD 1979a].

Oil

There is a trend toward reduction of oil-fired electricity-generating capacity in all countries addressed in this book, particularly in the United States.

Natural Gas

No new gas-fired electricity-generating capacity is planned for any of the countries addressed in this book.

Nuclear Power

Nuclear power is expected to assume a much higher proportion of electricity-generating capacity in the countries addressed in this book. This expectation may diminish in light of persistent and increased public opposition, commitment of substantial resources to new forms of energy, and public acceptance of large-scale energy-conservation programs [OECD 1979b].

INTERNATIONAL COOPERATION

International cooperation with respect to pollution control is strong within European countries, particularly those that belong to the Organization for Economic Cooperation and Development (OECD) and its International Energy Agency (IEA), the European Economic Community (EEC), and the Economic Commission for Europe (ECE). For a discussion of this subject, the reader is referred to Appendix B.

With respect to power plants, the purification of waste gas emitted to the atmosphere is of great importance internationally. Purification prevents sulfur dioxide emissions, which are released into high atmospheric layers from tall stacks and are subsequently transported to and deposited in distant areas. Recently, such long-range transport has received considerable international attention for its contribution to acid rain.

In 1979 member countries of the ECE signed an agreement to counteract long-range transboundary air pollution. A subsequent EEC directive for sulfur dioxide and suspended solid matter specified that no measures are permissible that would only shift ambient air pollution to regions that have relatively clean air [Scharer 1980].

The countries addressed here are members of several international organizations that sponsor studies and develop standards related to pollution control: the ECE (of the United Nations), EEC, IEA (of the OECD) and OECD. A discussion of the activities of these organizations and their effect on member countries with respect to pollution control is presented in Appendix B.

U.S. ELECTRIC POWER INDUSTRY

The electric power industry is the largest U.S. industry in terms of gross capital assets, and accounts for 10% of annual investments by U.S. industries. This prominence is due, in large part, to the versatility of its product (electricity), which affects all sectors of the U.S. economy, including industry, government and households [Bich 1977].

The industry encompasses about 3000 owners, which include private investors, federal agencies, state and local public agencies, and rural cooperatives. Investor-owned utilities account for 78% of the net generation of electricity in the United States, utilities operated by federal agencies (Tennessee Valley Authority and five federal power-marketing agencies) 10%, public nonfederal entities (municipal, special utility districts and state authorites) 9%, and rural consumer-owned cooperatives 3% [DOE 1981a].

In 1970 U.S. utilities accounted for 24% of total domestic fuel use. In 1980 this level increased to 32%. Until recently, industry estimates projected that electric utilities will account for almost half of total U.S. energy consumption by the year 2000. However, increased electric bills may force many customers to reduce electricity consumption by using alternative energy sources, such as gas furnaces, wood stoves, solar water heaters, district heating and various energy conservation devices and techniques [GAO 1981a].

The mix of fuels used to produce electricity has changed in the last two decades. This change reflects the cost and availability of fuels as well as changing technologies. For the last 30 years, coal has been the primary source of energy for electricity generation, accounting for about half of the electricity produced. During the 1960s, the remaining 50% was produced from oil, gas and hydropower. Since that time, nuclear power has grown, and by 1980 it produced 13% of U.S. electricity. Of the remaining noncoal power in 1980, hydropower accounted for about 10%, gas 12%, and oil, 15% [GAO 1981a].

Fossil-fuel electric power stations are considered to be major sources of air pollution, because they burn the most polluting fuels in the largest quantities. Of the fuels used, coal results in the most air pollution and natural gas in the least. In 1975 more than 65% of the coal used to produce steam in the United States was used for power generation. About 16% of all fuel oil and 10% of natural gas consumed also was used for the production of electricity. The oil used was mostly residual oil, which has a higher potential for air pollution than the more expensive lighter fractions, because of its higher sulfur and ash contents [EPA 1979].

The early 1970s were a financial watershed for electric utilities. The rate of technological advances that had lowered costs for the previous three decades was losing momentum and a number of environmental, institutional and fuel-related factors combined during this time to drive up the costs of new utility facilities.

The Clean Air Act of 1970 was the largest contributor to the increased environmental costs of power plants. Before 1970 few, if any, controls were required for air emissions; simple landfill disposal of fly ash was permitted; and condenser heat was rejected directly into water bodies via once-through cooling systems [Krohm 1980].

After 1970 environmental safeguards greatly increased the capital and operating costs of power plants. In the early 1970s these air emission controls added about 23% to the capital cost of a new baseload coal plant, and the Clean Air Act Amendments of 1977 increased this to about 29% of the capital cost. Cooling towers, widely used in new plants after 1970, increased the total capital cost by roughly 10%. Air emission controls and cooling towers also increased the operating and maintenance costs and decreased plant efficiency. Solid waste disposal was a minor problem before the time sulfur dioxide scrubbers came into use. By the end of 1980, the cost for disposal of scrubber sludge averaged $5–10/ton. About 3–4 tons of coal burned in a nonregenerative scrubber produces about 1 ton of sludge [Krohm 1980].

CHAPTER 2

POLLUTION CONTROL STRATEGIES

FEDERAL REPUBLIC OF GERMANY

Law and Government Structure

Germany is a federal state. The present German constitution (Grundgesetz), adopted in 1949, gives the 11 German states (Länder) equal status with the federal government (Bund). Legislative authority is divided between the federal government and the Länder in three ways. In some areas, such as national security, the federal government has exclusive jurisdiction. In other areas, such as control of air pollution, noise pollution, drugs, poisonous substances, waste, consumer protection and trade regulations, the federal government and the Länder have concurrent jurisdiction, but the Länder may only pass laws in these areas if the federal government has not done so. In the areas of nature protection, landscape protection, land use and water law, the federal government can only pass broad "framework" laws (Rahmenvorschriften). The Länder must implement the federal law by passing detailed laws adapted to the different conditions in each state [Bureau of National Affairs 1978].

The Länder, through their administrative agencies, are primarily responsible for enforcing most land-use and environmental laws, both federal and state. The federal government has a limited power to review state enforcement of federal laws. If the Länder are empowered by the constitution or by statute to administer federal law, they may be bound by directives issued by the federal executive branch, after consultation with the upper house of parliament (Bundesrat). These directives are sent to the Länder only, not to lower levels of administration [Bureau of National Affiars 1978].

Table 2-1. Major Environmental Protection Laws of the Federal Republic of Germany That Relate to Electric Power Plants

	Nature of Regulation
Air	
Federal Air Quality Control and Noise Abatement Act of 1974 (also known as the Federal Immissions Control Law)	Sets emission and immission (ambient air quality) standards
Third Ordinance under the Federal Air Quality Control . . . Act (1974)	Sets limitation on the sulfur content of fuel oil
1974 Technical Instructions on Air Pollution Control (TA Luft)	Requires FGD for new power plants
VDI Rules	Guidelines for air quality standards and their interpretation as determined by the VDI (Association of German Engineers)
Water	
Water Management Act of 1976	Establishes uniform water pollution control planning and management
Wastewater Charges Act of 1978	Establishes fees for different levels of polluted water discharged
Federal Air Quality Control . . . Act of 1974	Regulates thermal discharges and water discharges from cooling towers
Noise	
Federal Air Quality Control and Noise Abatement Act (1974)	Establishes standards for reducing sound from equipment in new facilities
Solid Waste	
Law on the Disposal of Wastes (1972)	Provides for management of solid wastes by the Länder
Federal Air Quality Control . . . Act (1974)	Regulates solid wastes resulting from incineration
Siting	
Federal Regional Planning Law (1976)	Controls regional siting of power plants by coordinated planning by the Länder
Federal Town Planning Law (1965)	Controls local siting of power plants by local, specifically public, involvement
Federal Atomic Energy Law	Controls siting and construction planning for nuclear power plants
Planning	
Third Electricity Generation Act (as amended in 1976)	Streamlines the implementation of major energy projects and focuses on use of coal as energy source
Federal Law of Environmental Statistics (1974)	Requires submission of data on air, water and solid wastes for assessment and planning purposes at the federal level

Table 2-1 is a listing of the major environmental protection laws in the Federal Republic of Germany together with the nature of their regulation of the electric power industry. These are discussed in detail in the following sections.

Air Pollution Control Strategy

Air quality standards and emission standards are implemented in the Federal Republic of Germany on the basis of the "Law for the Prevention of Harmful Effects on the Environment Caused by Air Pollution, Noise, Vibration and Similar Phenomena" (official translation—Federal Air Quality Control and Noise Abatement Act). This law is a continuation and modernization of the Commerce and Industry Act of 1869, especially with respect to air pollution control. Thus, it is based on more than 100 years of experience and tradition in regulatory control of industrial practices [Weber 1981].

Installations comprising a particular potential to cause harmful effects on the environment must be granted a license. For the licensing of such installations, the federal government has implemented air quality and emission standards by the First General Administrative Regulations under the act, dated August 28, 1974, as "Technical Instructions for Air Pollution Control" (Technische Anleitung zur Reinhaltung der Luft, known as "TA Luft"). Air quality and emission standards are of vital importance for installations subject to licensing. They form the basis for any administrative activity for all of the licensing procedures provided by the law, enabling plant operators to take account of them when planning and deciding on investments [Weber 1981].

The control strategy is based on a stringent system of preconstruction permits. Before a new plant may be built (or an existing one altered, or even an existing process altered), the permission of the Board of Works (Baurechtsamter) and the Trade Supervisory Authority (Gewerbeaufsichamter) must be obtained. This is similar to the new source review program under the Clean Air Act in the United States. The permit to each entity requires the installation of pollution control devices, corresponding to the current state of technology, that are economically feasible for that industry. This requirement is similar to the best practicable control technology requirement of U.S. legislation. Imposition of site-specific emission requirements is designed to reach attainment of air quality standards within the vicinity of the proposed facility. The standards were based on extensive studies and the development of a criteria document for each pollutant. These "immission" standards are shown in Table 2-2 [Mangun 1979; OECD 1975a,b; Weber 1981].

Table 2-2. Federal Republic of Germany Immission Standards[a] [Weber 1981]

	For Protection Against	
	Short-Term Effects	Long-Term Effects
Dustfall (mg/m^2-day)	650	350
Suspended Particles (μg/m^3)	300	150
Chlorine (μg/m^3)	300	100
Hydrochloric Acid (μg/m^3)	200	100
Hydrogen Fluoride (μg/m^3)	3	1
Carbon Monoxide (mg/m^3)	30	10
Sulfur Dioxide (μg/m^3)	400	140
Hydrogen Sulfide (μg/m^3)	20	10
Nitrogen Dioxide (μg/m^3)	300	80
Nitrogen Oxide (μg/m^3)		200
Lead		
Dustfall (μg/m^2-day)		500
Suspended Particles (μg/m^3)		2
Cadmium		
Dustfall (μg/m^2-day)		7.5
Suspended Particles (μg/m^3)		0.04

[a]Standards relate to "immission" levels, i.e., ambient air quality, usually in an urban area within the vicinity (within a few meters) of the emission source (proposed definition of the term "vicinity" may delimit an area 1 km square). These standards reflect the 1978 amendments to the 1974 TA Luft.

The permit to commence operation of a plant is also subject to objection by third parties. The permit application and the supporting documents (except trade secrets) must be subjected to public inspection for a specified period of time. During this period, interested parties are given an opportunity to make objections to the issuance of a permit. If objections are made, the licensing authority must discuss them with both the applicant and those who have lodged the objections.

The Air Quality Act requires states to develop "Clean Air Plans" for places where "harmful effects on the environment occur, or can be expected to occur, by air pollution within the whole or parts of areas with a heavy pollution load." This requirement is similar to the air quality management plans required under the U.S. Clean Air Act.

One aspect of the German air pollution control strategy that differs significantly from the U.S. strategy is the requirement that one or more immission control officers be appointed by the operators of installations subject to permitting requirements. The responsibilities of the immission control officer include:

1. to influence the development and introduction (1) of processes compatible with the environment, including processes for the proper use of residual

substances resulting from operations, (2) of products compatible with the environment, including reextraction and recycling processes;
2. to cooperate in the development and introduction of processes and products compatible with the environment, especially by submitting advisory reports on processes and products as to their environmental compatibility;
3. to ensure the observation of the provisions of the act and of the ordinances issued on the basis of the act as well as compliance with conditions and restrictions, especially by inspecting the works at regular intervals, measuring emissions and immissions, reporting deficiencies and submitting proposals to remove such deficiencies; and
4. to inform employees about the harmful effects on the environment caused by the installation and about the facilities and measure for their prevention, making allowance for the obligations arising out of the act or ordinances issued on the basis of the act.

Furthermore, the operators of the installation are to obtain the views of the immission control officer(s) before making investment decisions that may be significant for immission control. The opinions of the control officer must also be made known directly to the management if he is not able to reach agreement with the competent operational head [Mangun 1979].

Control of Power Plant Emissions

Sulfur dioxide is considered one of the major air pollutants. It exceeds the permissible short- and long-term standards in heavily polluted areas. In some urban areas, its levels are about 70 $\mu g/m^3$ of air, exceeding the 60-$\mu g/m^3$ standard recommended by the World Health Organization [UmweltBundesAmt 1980].

Power plants and heating stations, including industrial power plants, account for 50% of the total sulfur dioxide emissions in the country. However, in densely industrialized areas, power plants emit 40% of total emissions, but account for only 14% of the ambient sulfur dioxide in their vicinity. This is due to the high stack heights of the power plants, which promote the dispersion of emissions over long distances [UmweltBundes-Amt 1980].

The Federal Republic of Germany has spent considerable funds on promoting technologies that reduce emissions of sulfur dioxide. Its approach is the rational and economic use of energy, desulfurization of fuel oil, improved coal cleaning, flue gas desulfurization (FGD) in major installations, and increased use of gas and nonfossil fuels. FGD is the method of choice to obtain a marked decrease of emission levels [Federal Republic of Germany 1981; UmweltBundesAmt 1980].

The federal government provided the necessary prerequisites for these measures through the implementation of several initiatives. In the Third

Ordinance under the Law on Air Pollution Control, the sulfur content of fuel oil is limited to 0.3%. The effects of this ordinance are focused on small installations and road traffic, which are the major causes of air pollution in urbanized areas. With its program for the rational and economic use of energy, the federal government has provided the basic legal instruments and programs for energy saving [UmweltBundesAmt 1980].

Especially in concentrated areas, these measures will lead to a decrease in ambient air pollution through energy economy, particularly in private households and small-scale consumption, and to the shifting of emission from numerous minor coal-fired installations to a small number of large emitters, where stack gas purification techniques may be adopted. Since 1974, when TA Luft entered into force, major new power plants were licensed under the condition that they were equipped with FGD equipment. By 1980 three commercial FGD facilities were put into operation in new hard-coal-fired power plants. FGD was developed to the commercial stage through federal grants. As a result, it has become possible to meet the emission limit of 650 mg SO_2/m^3 waste gas, which was recommended at a conference of the environmental Ministers of the Federation and the Länder [UmweltBundesAmt 1980].

In 1978 the federal government adopted several legal measures to facilitate and harmonize the collection of air quality data. An ordinance on emission declarations requires the operators of specific types of facilities to submit detailed information to responsible authorities on all emissions released from a plant during the previous year. This emissions declaration is to be completed and updated annually. Emissions declarations are mainly prescribed in heavily polluted areas. For several types of facilities, however, emissions declarations also must be submitted if they are situated outside the polluted areas. This applies to larger emitters, such as power plants of 4 MWe and more [Weber 1981].

Water Pollution Control Strategy

The water pollution policy of the Federal Republic of Germany is based on the use of minimum effluent standards. These are to incorporate "best practicable means" in reducing effluent concentrations by "utilizing current available technology." These minimum standards must be met for all discharges. Federal guidelines for concentrations to be attained by effluent treatment exist, but in themselves have no legal status. The Länder governments apply these on a case-by-case basis when issuing license conditions [OECD 1979a,b].

In Germany, the Länder have the responsibility for issuing discharge permits. Even if they delegate this responsibility to a municipality or water association, the Länder still bear the ultimate responsibility of oversight and enforcement. Amendments made in 1976 to the Water Management Law of West Germany indicate that unlawful discharges can result in a fine of up to 100,000 deutschmarks or imprisonment, resulting in an economic penalty more severe than under U.S. law [Mangun 1979].

A major difference between the water pollution control strategies of the United States and West Germany is in the form of the effluent surcharges or taxes. Before January 1, 1978, the policy was not uniform throughout West Germany. Until this time, the Water Management Associations in North Rhine-Westphalia extracted an effluent discharge fee from each entity that discharged into their water basins. All of the money collected was used for clarification of degraded water and maintenance of an adequate supply of clean water for drinking and other domestic purposes. At the beginning of 1978 the Wastewater Charges Act (Abwasserabgabengesetz) went into effect and extended this policy uniformly across the entire country. The Länder implement the law and collect the charges (beginning in 1980) according to a set system. The amount of the charge depends on the noxiousness of the wastewater, which is determined on the basis of its volume, settleable solids, oxidizable substances and toxicity. All of these parameters are expressed in units of noxiousness [Mangun 1979].

One potential weakness in the act relates to charge liability. The federal government has the authority to exempt discharging parties until 1990. Such exemptions will not affect the conditions of the permit, which specify the maximum amount of polluted water permitted to be discharged annually for each discharger [Mangun 1979].

The federal government of Germany had hoped that the Water Management Law of 1976 would enable it to establish a more uniform program through centrally controlled efforts, as in the United States, but this did not come about. The Bundesrat, being composed of representatives from the states, refused to grant the powers necessary to the central government to give it the jurisdiction to uniformly regulate water law. The closest thing to uniformity to result from the law was the obligation of the states to administer water pollution control on the basis of "generally recognized rules of technology." This meant that the states were to adopt effluent limitations on the basis of "generally recognized rules" rather than according to the "state of technology," which the federal government wanted. These definitions are comparable to the U.S. concepts of "best practicable technology" and "best available technology" [Mangun 1979].

Because of strict constitutional limitations, the German federal government does not initiate enforcement actions in the face of failure by state governments to do so. However, since the responsibility lies clearly in the domain of the states, they have responded accordingly with active enforcement programs that promote a quick response to violations of the federal and state environment laws. All states except Hamburg have established special environmental units in their public prosecutors' offices to promote prompt prosecution of environmental offenders [Mangun 1979].

In Germany, an integrated river basin development planning program is in operation. As in the United States, the states are responsible for preparing the water management plans. These plans contain information needed to guide the use of all waters in the respective basins. However, it is not possible to observe gross geographic patterns for water quality throughout the country, as a national monitoring network does not exist. Trends in water quality currently can be perceived only in the particular river basins where studies have been conducted [Mangun 1979].

Control of Power Plant Discharges

In addition to federal and Länder laws relating to water discharges, there is a technical committee (Länderarbeitsgemeinschaft Wasser, i.e., Interstate Working Group on Water Problems) that has established the technical and physical maxima for the temperature of water bodies. The use of water by power stations is governed by concurring legal instruments [ECE 1981].

Usually, hydraulic models are used to conduct hydrothermal studies of specific cooling configurations. The models test technical design reliability, effectiveness of reducing temperature of the discharged water, verification of intake and discharge conditions, recycling patterns and rates, and effects on navigation [ECE 1981].

In Germany, the power plant operator decides on the type of cooling system for a proposed plant and seeks approval from the Länder or from local governments (Regierungspräsidien) to use a freshwater source for cooling purposes. There is a legal entitlement for power plants if they are determined to be needed in the "public interest." In 1973, 50% of all German thermal power plants used direct cooling as a method to dissipate heat, with the remaining 50% using wet cooling towers. By 1981, 75% of thermal power stations had cooling towers [ECE 1976a,1981].

A discharge license is usually granted for the lifetime of the power station; however, provisions exist for alteration of conditions or actual withdrawal of permission to discharge as permitted under the license. The procedure for application for a license to discharge involves submission of a petition for the right to discharge (i.e., a Wasserreichtsantrag), which is to provide a detailed description of the water flow required, quality of the water discharged, and the design of the intake and discharge structures. Also in support of the discharge application, model tests must be made relevant to transversal velocity of discharged water, temperature diffusion gradient and effects on navigability [ECE 1976a].

The temperature limit of the receiving body of water is 28° C after mixing, with a limit of temperature rise after mixing of 3° C (as of 1975). The limitation of increase in temperature of the condenser water is 10° C with a limit on the temperature of the cooling water before discharge of 30° C maximum (as of 1975). With regard to closed-circuit wet cooling, approval for a discharge permit also is required, but the limit of increase in temperature between intake and discharge is 15° C with a maximal discharge temperature of 35° C. Studies are being carried out on fish-farming in closed-cycle water-supply systems [ECE 1976a,1981].

One unique aspect of the effects of power plant discharges in Germany is that the organic pollution of rivers caused by incorrect use of water for domestic and industrial purposes is being reduced as a result of the presence of heated cooling water (capacity for self-purification increases). The self-purification process requires the presence of sufficient quantities of oxygen in the water. The fact that the oxygen content of the water discharged by modern power stations is higher than the natural oxygen content in heavily polluted rivers leads to an improvement in water quality [ECE 1981].

Noise Control Strategy

Regulation of noise within power plants is unique to the Federal Republic of Germany. It is a part of a national program (Federal Air Quality and Noise Abatement Act) to respond to the results of a survey by the Federal Statistics Office that showed that one in four local residents complain about noise nuisance. While most current regulatory effort is focused on traffic, industry is encouraged to install silencers on equipment, use sound-damping procedures, and to restrict construction to daylight hours [Jansen, 1981; Simek 1979].

Maximum noise limits have been set, and facilities are required to apply the "state of technology." If the cost of sound-proofing, such as construction of walls or earth barriers, is disproportionate to the effect, the builder may be exempted from the requirement, but then may have to compensate people disturbed by the noise. Construction noise from compressed-air hammers, cranes and other machinery is regulated by a series of federal regulations that set standards for ambient noise emissions, which are based on current technical standards for production of construction machinery and measuring techniques [Bureau of National Affairs 1978].

Solid Waste Control Strategy

Treatment and collection of waste is regulated by the Law on the Disposal of Wastes (1972), which prohibits the disposal of waste in any way that endangers human health or welfare or damages the environment. In compliance with the law, the Länder have enacted specific "implementation laws" for waste disposal [Bureau of National Affairs 1978].

The Federal Air Quality Control Act (1974) limits, in connection with special regulations, emissions from waste disposal activities with respect to incineration and composting. According to this law, new industrial facilities only are allowed to be constructed if their wastes can be treated properly [OECD 1976a].

Another law that has importance with respect to the handling of power plant wastes is the federal Law of Environmental Statistics (1974). This law requires the collection of data concerning the quality and quantity of wastes [OECD 1976a].

Power Plant Siting Strategy

The Federal Republic of Germany comprises 11 Länder (regional governments) of varying size and population density. Several (Bremen, Berlin, Hamburg, Westphalia and Saarland) have the highest population densities in Europe [OECD 1980].

Most federal laws are implemented by the Länder as if they were Land (regional government) laws. On behalf of the federation, the federal government enforces specific laws and enacts general administrative procedures to ensure uniform implementation of federal laws by all Länder. The Länder can promulgate regulations, but they must not be in conflict with those of federal law. In special cases, e.g., the supervision of nuclear power plants under the Federal Atomic Energy Law, federal supervision covers not only legality, but also enforcement [OECD 1980].

Thus, licensing of new power plants is a function of the Länder. The Länder review norms for concentrations of pollutants in air and water discharges from power plants, amounts and quality of solid wastes, relevant air and water quality standards, and nature of proposed discharges, and consult with various national agencies. Nuclear plants require more stringent review. In addition to local authorities, licensing also occurs within the federal Ministry for the Interior [OECD 1977].

The two main federal land-use laws are the federal law concerning Regional Planning and the federal law for Town Planning. The goals of these laws are to protect human rights within a community while allowing for economic, social and cultural development of the country. The laws provide for the development of regional and local land-use plans and for active public participation in public hearings.

Most land planning laws refer to siting of major energy facilities and protection of the environment. In most cases the emphasis is on security of electricity supply at a reasonable price. There is, however, a trend toward increasing protection of the environment. Given the increasing difficulty of finding sites for major energy facilities acceptable by the local population and the Länder governments' concerns for meeting the demand of energy, Länder planning authorities have placed considerable emphasis on solving siting problems [OECD 1980].

A license is required for constructing and operating electric power plants. The licensing procedure for nuclear power stations is defined by the Federal Atomic Energy Law, and that for conventional power stations and other major energy facilities by the Federal Air Quality Control Law. In both cases, one of the licensing requirements is that the site conform with the provisions of the federal Town Planning law.

The Federal Atomic Energy Law is applied and enforced by the Länder on behalf of the federal government. The law stipulates that the land licensing authority for nuclear energy facilities should be at the highest level (ministry). Federal, Land and community statutory authorities affected by the application take part in the licensing procedure. The procedure followed is that of the federal Immissions Control Law.

According to the federal Immissions Control Law, a license to construct and operate a major energy facility may be granted only if it has been shown that:

1. harmful environmental impacts for the public in general and the local community in particular will not arise;
2. measures for the control of harmful environmental emissions will be taken, using the best available technology;
3. waste produced by the facility will be recycled and if this is technically or economically impossible, waste will be disposed of in accordance with existing regulations; and
4. other public laws or worker protection rules do not stand in the way of the construction and operation of the facility.

To promote the development of domestic energy resources, Germany amended (in 1976) its Third Electricity Generation Act. The aim of the amendment was to reduce dependence on oil and to increase the use of coal and nuclear energy for power production by implementing a new federal program of energy independence. The program places special emphasis on the need to acquire sites for power plants. The 1976 federal government report on the environment summarizes the provision under federal law to obtain sites for major energy facilities. The report indicates that

> suitable sites for industrial facilities of more than regional importance, particularly major energy facilities, must be duly identified. In cooperation with the Länder, the federal government will work out a scheme for planning and securing sites, including supply and disposal facilities in the energy field, within long-term regional development plans with the participation of the affected population. This work program comprises general requirements and assessment data for an early examination of possible sites, a survey of land-use plans, agreements on projects near to borders between Länder or with neighboring countries, and a long-term siting policy within EEC.

The provision laid the cornerstone for the development of a new land-use planning approach that permitted the Land planning authorities to ensure the smooth implementation of major projects of political and economic importance by influencing community land-use planning. Thus the Land government is now able to use the recently acquired legal machinery for siting large industrial projects, including power stations [OECD 1980].

Public Apprehension to the Siting of Power Plants

There is an increasing opposition to new locations of power plants on the part of individual groups and occasionally on the part of the authorities. In some cases the opposition of the public to power plants, big cooling towers and overhead high-voltage lines is so great that the authorization is refused for a power plant that is planned or under construction and for planned routes of transmission lines. Instead of overhead high-voltage lines, cabling is increasingly required, particularly for towns and recreational areas. However, in general, cabling is rejected by the authorities because ease of access is less than that of overhead lines, and because from an overall economic point of view, cabling is unacceptable [ECE 1976b].

The public has been highly interested in the restriction of air pollution by power plants. This concern has been alleviated by the passing of appropriate legislation, namely the federal Immissions Control Law of 1974.

Objections have been raised to the construction of wet cooling towers because it is feared that they may influence climatic conditions. This will probably be the case only when many large towers are concentrated in a small area. Such a concentration is not being planned. Noise caused by cooling towers is often felt to be a drawback [ECE 1976b].

The maximum allowable radioactive emissions from nuclear power plants are prescribed by law. In case of normal operation these limits are never reached. German experience shows that human beings living in the surroundings of a nuclear station are exposed to a radiation load exceeding the normal load by 1 mrem/yr, which is negligible compared with the normal load, which amounts to 100–120 mrem/yr. Still, people continue to voice their fears of nuclear power plants [ECE 1976b].

Perceived unreasonable public apprehension aroused by individual pressure groups is one of the main difficulties that must be overcome by German public utilities within the scope of their authority. Most objections (98%) are voiced by petitions circulated by organized groups. Public utilities are therefore increasing their public relations activities to convince the public of the low risk and of the fact that nuclear power stations do not pollute the environment. Information centers in each power plant provide information on this subject. The results of these increased public relations activities are noticeable [ECE 1976b; OECD 1977].

FRANCE

Law and Government Structure

Environmental protection legislation in France dates back to an imperial act of October 15, 1810. Its principles were brought into practice in other European countries with the influence and conquests of Napoleon. Deeply modified in 1917, 1932 and 1961, this legislation recently has been modernized and codified into a single all-encompassing act: the Installations Registered for the Purposes of Environmental Protection Act of 1976 (and the implementation decree dated September 21, 1977). The act brings together provisions of previous acts, namely, the Act on the Control of Atmospheric Pollution and Odors (1976), the Act on the Regime Governing Waterways and Water Distribution and Control of Their Pollution (1964), the Act on Waste Disposal and the Recovery of Materials (1975), the Act on Nature Protection (1976), and various decrees dealing with the imposition of charges applicable to establishments classified as dangerous, unsanitary, noisy or noxious [Service de l'Environnement Industriel 1977,1979].

The act is implemented by the Direction de la Prevention des Pollutions of the Service de l'Environnement et du Cadre de Vie. Within the broad organizational framework and guidelines established by the Ministere de l'Environnement et du Cadre de Vie, actual decisions on specific plants and issuance of permits are carried out under the authority of local prefects. Prefects govern departments (similar in area to counties within the United States). France is divided into 95 departments [Service de l'Environnement Industriel 1979].

The Ministere de l'Environnement et du Cadre de Vie maintains a list of installations and activities of particular industries that must be regulated to ensure that their operation will not entail pollution or substantial danger to human health and the environment. Plants on the list must comply with technical regulations that specify limitations on the release of air and water discharges and measures to prevent fires, explosions or accidental discharges [Service de l'Environnement Industriel 1979].

The main activity associated with the 1976 act is actual determination of technical requirements to be imposed on a facility that requests an operating license. When a request is made, the response is not a choice between acceptance or refusal. The regulatory response is to determine the precautions necessary to ensure that the facility will not be a source of unacceptable pollution or danger. Technical analyses performed mostly by the applicant play an important role in the decision-making process.

The permit itself is recognized as a charter for a permanent relationship among the environment, the administration and the manufacturer. Thus, the decision regarding permitted pollution to be released from a facility becomes an arbitration among many considerations, such as environmental protection, creation and protection of jobs, visual and auditory amenities of the neighborhood, reasonable costs, and industrial development [Service de l'Environnement Industriel 1979].

The responsibility for negotiating permits lies with the prefect. The act recognizes that industrial pollution control must take into account different points of view. Such decentralization of the decision-making process allows for discussion and cross-examination at a level where all of the various factors can be assessed clearly and publicly. It avoids federal permit negotiations, which usually are out of context and slow, and can be prejudicial to the economic aspect of a project without any profit to the environment [Service de l'Environnement Industriel 1979].

Issuance of a permit by the prefect establishes the technical requirements for the operating conditions. As a general rule, the kind of pollution control equipment to be used is not specified, but left to the choice of the operator of the facility. However, the permit does specify minimum controls required for plant safety.

The act clearly endeavors to promote waste prevention rather than waste treatment. Emphasis in its implementation, especially for new facilities, is to promote the use of internal modifications to the process that can allow the facility either to prevent polluting emissions or to recycle them [Service de l'Environnement Industriel 1979].

Air Pollution Control Strategy

In France, no industrial installations in service or under construction in 1977, including electric power stations, employed FGD.

In 1972, when conventional thermal power stations were expanding, the government required Electricité de France (EDF) to consider the provision of a FGD installation for a 250-MWe unit at its Martigues power station. This would have been the first step toward the general use of this means of protection for all new plants to be brought into service. A thorough study was carried out on that occasion after consultation with all potential national and foreign suppliers. There was found to be no available process that was industrially acceptable (in terms of sufficient guaranteed reliability, absence of harmful by-products, etc.). The installation had to be postponed, and at the same time France decided to intensify studies and research on the subject [ECE 1977a].

The oil crisis at the end of 1973 brought about a complete change of perspective. Electricity production, destined to be based mainly on nuclear power, is expected to not represent an atmospheric pollution problem after 1985. In view of the time required to develop prototype apparatus for FGD, it is unlikely that a substantial amount of the control equipment for electric power stations will ever come into existence in France [ECE 1977a].

Water Pollution Control Strategy

In France, the plant operator, usually EDF, decides on the type of cooling system for a proposed power plant. Permission must be given by the owner of the water for use of a freshwater source for cooling water. There is no legal entitlement; however, it is possible to obtain a license for use of the water. The licensing period is prescribed, usually the lifetime of the power plant, with provisions for amending or modifying the license. The procedure for acquiring a license varies according to the owner of the river or other waterway [ECE 1976a].

Temperature limits of the water body that receives the effluent are not fixed and there is allowance for an increase in temperature of the receiving water. However, no toxic or noxious compounds are to be discharged.

Preconstruction baseline ecological studies are performed based on the conceptual or general design of a proposed plant. These studies are made by the Direction Etudes et Recherches of EDF. Engineers also are employed to examine the proposed site and prepare a required site report [ECE 1976a].

For existing and planned power stations with natural-draft wet cooling towers, a number of special investigations and reports are required. These include studies of the behavior of the towers under varying meteorological conditions. The studies are reviewed, in compliance with the "Installations Registered" act of 1976, by the prefect, who may require an environmental impact assessment. The assessment, if performed, should take into account the water management plans of regional water authorities [OECD 1976b].

Noise Control Strategy

Noise control is required, if deemed necessary, by the department prefect in the conditions of the pollution control permit. Emphasis is placed on protection of the auditory amenities within the vicinity of the plant. Noise is not considered a major problem with respect to the operation of power plants.

Solid Waste Control Strategy

Solid waste from power plants also is regulated under the 1976 act. The conditions for disposal are determined by the department prefect in the issuance of a pollution control permit. Emphasis is on striking an optimum balance among process modifications to minimize solid waste generation, reduce air and water pollution, find beneficial uses for the solid wastes, and require safe disposal in a certified landfill.

Power Plant Siting Strategy

In France, the responsibility for determining suitable sites for the construction of power plants is held by EDF, the national electricity supply body, within the framework of its general functions of power generation, transport and distribution. The first step in the determination involves the search for "possible" sites. This is done mainly with the support of maps and local documents, with minimum site visits. At the end of this operation, there is a list of "possible" sites to which EDF would like access for a fixed period of time in the light of its regional consumption forecast, for government department examination at inter-ministerial meetings.

A few of these sites are then selected, each government department using criteria relevant to its own sphere. Thus, the meetings are generally attended by agencies having particular interest in the development of electric power in an environmentally acceptable manner. These are [OECD 1980]:

1. Delegation a l'Amenagement du Territoire et a l'Action Regionale (Regional Planning Committee), currently attached to the Ministry of Equipment and Regional Planning;
2. La Direction du Gaz, de l'Electricite et du Charbon au Ministere de l'Industrie (Gas, Electricity and Coal Directorate), Ministry of Industry;
3. Le Service Central de Surete des Installations Nucleaires (Central Service Department for the Safety of Nuclear Installations), Ministry of Industry;
4. Le Service Central de Protection contre les Rayonnements Ionisants (Central Service for Protection against Ionizing Radiations), Ministry of Health;
5. La Direction de l'Amenagement Foncier et de l'Urbanisme (Directorate for Building Planning and Urbanisation), Ministry of Equipment and Regional Planning; and
6. Ministere de l'Environnement et du Cadre de Vie (Ministry of the Environment).

At the end of this stage, the government has a list of "possible" sites agreed to a priori by all ministerial departments. The next stage consists of consulting the elected representatives and local authorities, to prune the initial selection and at the same time to conduct "preliminary" studies aimed at assessing the technical and economic "feasibility" of the project and its acceptability in terms of safety, radiation protection and environmental impact. Once the regional (or department) prefect has been instructed by the government to start the regional (and/or department) council consultation stage, aspects of the project or projects concerning the region or department are discussed [OECD 1980].

The discussion covers all regional and local implications of the siting of one or more power plants, at economic, financial, population (including construction labor problems), agricultural, ecological, etc., levels. Government and EDF experts intervene in the discussions, dealing in detail with the various aspects referred to above. In some cases representatives of nature-protection associations that are sufficiently influential and reputable to be regarded as "valid" spokesmen are invited to present their points of view. These discussions are not considered as preliminary work toward a decisive vote on the future of the project(s) examined. Neither the regional councils nor the general council hold any legal right to restrict the government's freedom of action regarding site selection. These discussions are considered by the government as a suitable means of consulting regional and local bodies with a view toward determining their preferences [OECD 1980].

The outcome of the siting discussions and decision is the transmittal of the file and application to the regional prefect for licensing the power plant in the public interest. The prefect examines the file and circulates the application among concerned public and private organizations within the region and department prefects for review and comment. EDF replies to the comments raised in this review. An indepedent review is made by the Industry and Mining Interdepartmental Service. The service is responsible for seeing that each department is consulted at the regional and local level. At the end of the review process, all comments, including the results of an environment impact assessment (usually required), are sent to the Council of State for review. The Prime Minister makes the ultimate siting decision based on the inquiry process [OECD 1980].

The siting process discussed above is highly fluid and is likely to change substantially as the Installations Registered for Purposes of Environmental Protection Act is implemented. The focus of the act is to place a greater responsibility for ultimate decision, with respect to environmental matters, in the hands of the department prefects. For the most current and comprehensive review of the siting procedures for power plants in France, the reader is referred to the literature [OECD 1980].

Public Apprehension to the Siting of Power Plants

In 1972 EDF had to revise its plans for construction of new power plants for the first time. In view of public concern over an expected increase in air pollution, authorities forced EDF to locate a 700-MWe unit, the site of which was planned in the Mediterranean area, in the western part of France. As a result of this decision, the unit was put into operation at a later date than planned (in 1976) [ECE 1976b].

With regard to nuclear plants, public concern has focused on the safety of new facilities. This has led to delays in siting and construction. In 1974 delays in the approval of a preliminary safety report by authorities forced EDF to order two 900-MWe pressurized water reactors of an already approved type at an earlier date than had been expected. As a consequence, a high-voltage line (two circuits of 380 kV) between the cities of Fessenheim and Bugey and the Paris area had to be constructed before 1978. At the same time, EDF ordered 900- and 1200-MWe units of a new reactor type that were to be commissioned in 1979 and 1980 at an earlier date to have the necessary time for accomplishing the siting procedures [ECE 1976b].

UNITED KINGDOM

Law and Government Structure

Responsibility for protection of the environment lies primarily with the Department of the Environment, which also monitors and, to some extent, controls housing and local government. These functions are carried out by Welsh, Scottish and Northern Irish Offices. Each has a separate Secretary of State. These departments have responsibility for the introduction and implementation of Acts of Parliament and statutory instruments, which are orders in council and regulations introduced by ministers through powers delegated to them by Parliament [Bureau of National Affairs 1980].

Within the Department of the Environment is a control unit on environmental pollution, containing a team of scientists and administrators that also coordinates UK participation in international pollution control activities. Other directorates within the department deal with noise, clean air, wastes, water and water engineering. The Royal Commission for Environmental Pollution was set up in 1970 as an independent body to advise the government on issues of national and international concern. The Clean Air Council reviews the effectiveness of legislation to control and abate air pollution, advising the Secretary of State on air pollution problems [Bureau of National Affairs 1980].

Local governments in England, except for Greater London and Wales, were completely reorganized in 1974. Forty-five first-tier or county councils were set up in England and eight in Wales. These councils have responsibility for long-term planning, including land use, highways and transport. Environmental health is the responsibility of district councils. In addition, there are eight metropolitan areas where the district councils have powers normally vested in county councils. Reorganization in Scotland also is based on a two-tier system that includes 9 regional councils, 53 district councils and 3 island areas [Bureau of National Affairs 1980].

Central to the problem of protection of the environment from the effects of fossil-fueled electric power generation is the work of the Central Electricity Generating Board (CEGB). CEGB is responsible for developing and maintaining an efficient, coordinated and economical system of supply of electricity in bulk for all parts of England and Wales. Also, it is required to assure environmental protection, including protection of material beauty, wildlife, and objects of architectural or historic interest. The United Kingdom Atomic Energy Authority (UKAEA) designs and constructs some nuclear power stations, but most are owned and operated by CEGB.

Table 2-3 lists the major environmental protection laws in the United Kingdom together with the nature of their regulation of the electric power industry. These are discussed in detail in the following sections.

Air Pollution Control Strategy

In general, Britain's air pollution control strategies are based on "best practicable means" to comply with industry-specific standards. They rely more on persuasion and cooperation than on enforcement by legal sanctions and more on a case-by-case partnership between industry and government than imposition of blanket government-generated rules. Standards are responsive to individual situations and to local variations in the capacity of the environment to absorb pollution without harming public health. Each emission situation is judged on its merits. Polluters are responsible for minimizing and preventing pollution by best practicable means, technologically, economically and environmentally. Major industrial polluters must submit an annual report to the appropriate government agency on the best practicable means they employ to minimize emissions [Congressional Research Service 1981].

Emissions to the atmosphere from thermal power stations are subject to the alkali, etcetera Works Regulation Act of 1906 and subsequent orders issued under the act. The first Alkali Works Act, issued in 1863, established the Alkali Inspectorate, which is responsible for oversight of the act and the more recent Clean Air Acts of 1956 and 1968. Air pollution is controlled centrally by the Alkali and Clean Air Inspectorate in England and Wales, and the Industrial Pollution Inspectorate in Scotland.

The 1956 UK Clean Air Act provides a legal definition of the term "practicable" as reasonably practicable regarding, among other things, local conditions and circumstances, financial implications, and the current state of technical knowledge. The financial implications refer to equipment that was technically feasible, but would be judged impracticable if costs were so high that the facility would be rendered unprofitable [OECD 1979a,b].

CEGB defines proposals for air pollution control at power stations. These are discussed with and agreed to by the Alkali and Clean Air Inspectorate, which represents regulatory authority, but the initiative is with CEGB. The current aim of CEGB is to diminish public complaints by a policy of using single multiflue tall stacks, careful siting, careful combustion control, efficient dust collection and, in special cases, conversion of power stations to use more suitable fuels.

Table 2-3. Major Environmental Protection Laws of the United Kingdom That Relate to Electric Power Plants

	Nature of Regulation
Multimedia (Including Noise)	
Alkali Works Act of 1863	Established the main national regulatory body (Alkali Inspectorate) for controlling air and water pollution
Control of Pollution Act of 1974	The main act designed to protect the environment; regulates solid wastes, air emissions, water effluents and noise; provides for its implementation at the district and local levels
Radioactive Substances Act of 1960	Controls low-level discharges of gaseous, liquid or solid radioactive substances
Electricity Act of 1957	Created the CEGB and established guidelines for protecting the environment from the effects of electricity generation.
Air	
Alkali, etcetera Works Regulation Act of 1906 and subsequent Orders	Controls emissions from power plants under the jurisdiction of the Alkali Inspectorate.
Clean Air Acts of 1956 and 1968	Determine control by the use of "practicable means" for air emissions (smoke, grit, dust, and combustion fumes) and are administered by the Alkali Inspectorate
Public Health Acts of 1936 and 1969	Established local controls over the height of power plant chimneys and the monitoring of "noxious or offensive emissions"
Air Quality Directive of 1980	European Economic Communities air quality guidelines, to which air pollution control efforts in the United Kingdom are focused
Water	
Water Act of 1973	Established regional water authorities responsible for water management
Solid Waste	
Nuclear Installation Acts (1959, 1965, 1969)	Regulate the handling and disposal of high-level radioactive wastes from power plants
Siting	
Town and Country Planning Act of 1971 (1972 in Scotland)	Places authority for land use planning and siting decisions with local authorities

The Alkali Inspectorate does not operate with any fixed financial rule of thumb; rather, it relies on its own experience and that of other facilities in the same line of business when assessing a particular plant's ability to pay for pollution control equipment. Also, the inspectorate does not approach pollution control purely by setting rigid emission standards (or acceptable emission levels). Acceptable levels, with regard to the circumstances, set by the inspectorate are used in conjunction with process specifications and methods of operation to formulate a standard to which registered processes are required to conform. The inclusion of process specifications in the approach implies an obligation to use certain methods, but whenever possible the polluter is left free to choose between methods of abatement that are acceptable to the inspectorate. The current emission limit established for new coal-fired power plants by the inspectorate is 0.115 g/m^3 for total particulate matter [Flowers 1976; OECD 1979].

The Clean Air Acts refer specifically only to smoke, dust and combustion fume. "Grit" is defined as particles above 76 μm in diameter; "dust" (not specifically defined) is generally said to refer to particles of 1–76 μm and "fume" to be less than 1 μm. Smoke can be "dark smoke" or "black smoke," according to the Ringelmann scale, a visual measure of smoke density [Congressional Research Service 1981].

After an industrial process (e.g., electric power) is registered, the Inspectorate, together with the applicable trade association and/or the individual plant operator, discusses the problem and possible remedies. The industry conducts the necessary technical investigations and research, which occasionally is subsidized by the government. On completion of this often lengthy process, the Inspectorate is provided the findings, prepares "Notes on Best Practicable Means," and sets emission standards, again in close consultation with industry. The standards specify emission levels that are considered to be currently achievable, given the available technology, the nature and effects of the pollutants, and abatement costs to industry. Standards are tightened as improvements in the production process and control technology permit. Pollutant dispersion and dilution are considered the principal means to render "unavoidable emissions . . . harmless and inoffensive," and to maintain acceptable ground-level concentrations. The Inspectorate, therefore, specifies stack heights as part of the issue of "best practicable means" [Congressional Research Service 1981].

Other acts also relate in some way to air pollution from power plants. The Control of Pollution Act of 1974 authorizes local authorities to conduct investigations of emissions from large facilities and to make the data available to the public. The act also gives the Secretary of State authority to control the sulfur content of fuel oil. However, most of the air pollution section of the act is not directed to the regulation of industry, but of emissions from motor vehicles [Bureau of National Affairs 1980; Department of the Environment 1978].

The Radioactive Substances Act of 1960 regulates low-level radioactive wastes derived from nuclear power stations. The act seeks to reduce the average exposure of the whole population of the United Kingdom far below the limit of 1 rem/person in 30 yr.

High-level radioactive wastes from nuclear power stations are regulated by the Nuclear Installation Acts of 1959, 1965 and 1969. The 1969 act imposes absolute liability for damage on the plant operator subject to a limit of £5 million. Costs in excess of this compensation are borne by the Secretary of State [Department of the Environment 1978].

The Public Health Acts (1969 most recent) are enforced by local authorities to control noxious or offensive emissions not otherwise regulated under other legislation. The acts do not deal with specific processes, but are focused on the protection of human health, regardless of source. In this light, local authorities also control grit and dust abatement from new furnaces, and are authorized to monitor emissions and ground-level concentrations, and to require plants to measure and record their grit and dust emissions [Congressional Research Service 1981; Department of the Environment 1978].

Because of the cooperation of government and industry, formal enforcement proceedings to control air pollution are rare. Without specific limits of contamination, and given the availability of subsidies and tax incentives, compliance in some form is simpler and less controversial. This approach, which assumes that air pollution is bad and that any reduction of pollution levels at any time is likely to be beneficial to society, obviates the need for drastic enforcement measures [Congressional Research Service 1981].

Within the purview of the national government, the Inspectorate plays an important role in regulating power plants. The height of power plant chimneys is dependent on local circumstances, primarily the need for adequate dispersion of gaseous pollutants, particularly sulfur dioxide. However, there is no direct method for local authorities to require specific controls over actual emissions or regulation of the chimney height, other than by using the general nuisance provisions of the Public Health Acts of 1936 and 1969 [Commission on Energy and the Environment 1981].

Water Pollution Control Strategy

In contrast to environmental protection problems related to air pollution from power plants, water pollution problems facing the electricity generating industry in the United Kingdom in past years were considered minor. A temperature rise of about 10° C across condensers has long been adopted for Rankine cycle steam plants. With natural water temperatures seldom reaching 22–24° C, and generally being much lower, approval for discharge of cooling water at the commonly chosen maximum of 30° C does not impose severe economic hardship. However, once such a temperature is established, the cost of lowering discharge temperatures by only a few degrees can cost CEGB several million pounds per year [Leason 1974].

Until April 1974, approvals to discharge cooling water into a river were granted by the appropriate River Authority (created by some combinations of River Boards by the Water Resources Act of 1963) in accordance with the Rivers (Prevention of Pollution) Act of 1951. As a result of the Water Act of 1973, the 27 River Authorities were regrouped into nine English and one Welsh regional water authorities, which in 1974 took over responsibilities from the River Authorities in addition to new responsibilities, such as provisions for sewage drainage and disposal. All of these Regional Water Authorities extend their jurisdiction into tidal waters and coastal seawater to control marine pollution. Their efforts are coordinated by a National Water Council comprised of members of the Authorities and appointees of the Secretary of State [Bureau of National Affairs 1980; Leason 1974].

Although a focus of controversy, warm water is not considered as "poisonous, noxious, or pollution matter" under the Control of Pollution Act of 1974. The stand of the electrical generation industry is that, at worst, the thermal effluents can affect the distribution, but not the kinds of species inhabiting waterways adjacent to power plants. In some instances, it has been demonstrated that thermal effluents cause an increased growth rate of fish, prawns and oysters living in power station cooling water discharges [Leason 1974].

In the United Kingdom, the use of water by thermal power stations and the maximum temperature of water discharged by them are regulated by the same laws that govern the intake and discharge of water by other industries. The main regulatory acts are the Water Act of 1973, which established the aforementioned management structure implemented by water authorities, and the Control of Pollution Act of 1974, implemented jointly by the water authorities and the Alkali Inspectorate. The Water Act focuses on the management of water quantity. The Control of Pollution Act focuses on the management of water quality (and does not use fixed effluent standards).

In England and Wales, intake and discharge of surface water is controlled without the use of fixed standards by the regional water authorities responsible for the use and protection of water resources.

Decisions concerning the intake of water and the discharge of wastewater are made by local authorities responsible for the use and protection of local water resources. These local authorities determine the conditions governing the discharge of wastewater into water bodies in the light of its nature, composition, temperature and volume and the rate of discharge [ECE 1981].

There are no uniform legislative measures in force to regulate the use of water by thermal power stations in Scotland and Northern Ireland. For example, the maximum permissible temperature of wastewater may vary depending on local conditions, the properties of the water, or the purposes for which the water bodies are used. Water use is determined by the independent authorities responsible for the management of each water basin [ECE 1981].

Currently, about 45% of thermal power stations in the United Kingdom use once-through cooling water systems. About 48% use closed-cycle systems with cooling towers. The remaining 7% use combined once-through cooling with cooling towers [ECE 1981].

Noise Control Strategy

Noise from power plants is regulated by local authorities. The two approaches to control are: (1) use of best practicable means available, and (2) the establishment of a decibel limit at a defined distance from the facility. The definition of "best means" is difficult to define and for industry to achieve. Most local authorities have set a decibel standard and have defined site-specific noise abatement zones. Such noise control is authorized under sections dealing with noise in the Control of Pollution Act of 1974 [Barrett 1982; Department of Environment 1978].

Solid Waste Control Strategy

CEGB is responsible for disposal of about 10 million ton/yr of ash (1974), representing 6% of the total of mining, quarry, ash and clinker wastes in the country. Commercial uses absorb about half the amount [Department of the Envirnment 1979; Leason 1974].

Disposal of waste is under local control, at county council level through waste disposal departments, which also have the responsibility for licensing under the Control of Pollution Act of 1974. In practice, however, this responsibility is shared with the water authorities, who must be consulted on the effect of any proposed site on water supplies in the vicinity. The water authority may appeal to the Secretary of State if there are differences of opinion as to potential hazards [Bureau of National Affairs 1980].

Ash has been used to restore the natural levels of areas scarred by surface excavations and to reclaim land on estuarial boundaries. Such disposal schemes have been approved by local authorities, usually at the planning stage of the power station [Leason 1974].

Under the Control of Pollution Act, power plant solid wastes must be deposited at waste disposal sites specifically licensed to receive such wastes or at other sites under conditions determined by the authorities. Consequently, the primary method of establishing control over waste disposal operations is site licensing, whereby operators of proposed plants are required to obtain a site license from the local waste disposal authority [Department of the Environment 1978].

Power Plant Siting Strategy

Production and transmission of electric power in the United Kingdom is almost exclusively in the hands of the nationalized industry, i.e., CEGB for England, Wales and the south of Scotland, and the Hydroelectric Board in Scotland. Any proposed new plant, fossil- or nuclear-fueled, must receive the consent of the Secretary of State for Energy (or Secretary of State for Scotland) who is the minister responsible for verifying the need for a given power plant [OECD 1977].

Siting decisions are influenced by the 10 regional water authorities as they implement various sections of the Pollution Control Act of 1974. Local authorities also need to be consulted in compliance with the Town and Country Planning Acts of 1971 (England and Wales) and 1972 (Scotland). The local planning councils are responsible for long-range land-use planning. When the local planning authority considers a proposed siting plan to be a substantial departure from the local land-use plan, the local authority must give the public an opportunity to be heard through public inquiries or written responses. Under such conditions the local authorities also must inform the Secretary of State who can choose to review the application in light of the objections raised by the local authority [OECD 1977].

Public Apprehension to the Siting of Power Plants

Within the last few years as various provisions of the comprehensive Pollution Control Law were being implemented, public concerns relating to noise, air pollution, water pollution and impact on scenic vistas have been addressed. CEGB has had an integrated environmental control policy since its formation. Such a policy is not entirely without return, as it must reduce justifiable opposition to new projects. Care during the planning stage gives architects and landscape consultants the maximum opportunity, in close liaison with project management, to influence the layout and appearance of new power stations and to seek views on the proposals from independent bodies such as the Royal Fine Arts Commission [Leason 1974].

Major power plant amenity improvements include metal-clad switchgear and the assisted-draft cooling tower. The latter is entirely an amenity development aimed at reducing the number of cooling towers on a site, and its adoption increases the cost of cooling. Even broader interpretations of the act include a provision of CEGB land for nature trails and augmentation of game fishing on associated rivers and lakes [Leason 1974].

Opposition to proposed power station sites has been increasing since the 1970s with the general upsurge in public concern as to anticipation of pollution and loss of amenity. If the land for stations must be acquired compulsorily, CEGB must justify this to the government. When there are objections, a public inquiry is held. Sites, especially those near or inside population centers, are usually contested on grounds of anticipated pollution effects. The growth of amenity "pressure groups" promoting sectional interests and increasing general public concern for the environment have promoted a greater willingness on the part of the government to order public inquiries, which recently have tended to be lengthy and complex [ECE 1976b].

As to sites outside large population centers, the opposition often takes the form of protest against the loss of visual amenity, e.g., the intrusion of industry into a predominantly rural setting. Such opposition normally is expressed through national amenity societies or local resident associations. Marked opposition to transmission lines centers on degradation of visual amenity. Most major proposals in recent years have entailed a public inquiry in which the route proposals were subjected to a detailed examination [ECE 1976b].

Some sites give rise to considerable opposition because of the anticipated air pollution. The public mainly is worried about sulfur dioxide emissions (and dust emissions in the case of coal-fired stations). Control measures taken by CEGB ensure that power stations can have no harmful effects. Although CEGB has sponsored considerable research on air pollution control, opposition has remained strong, probably as a result of widespread publicity on pollution in general. This opposition is not likely to diminish in the next few years. The objection to existing stations is restricted almost entirely to emissions from the older power stations in urban areas [ECE 1976b].

Since natural water supplies in Britain are limited, power stations on inland sites must use cooling towers. Opposition to cooling towers has taken two forms: (1) objections to the bulk of the towers and their visible plumes, and (2) concern about possible effects on local climatic conditions, such as rainfall, humidity and sunshine. The effects on sunshine have been found to be negligible in practice, although objections related to this problem may increase in the future. It is anticipated that the visual impact will be decreased by the development of assisted-draft cooling towers, each of which can replace three or four natural-draft towers of the same chimney dimensions. The use of expensive dry-cooling techniques solely to avoid visible plumes is not considered to be justified [ECE 1976b].

There has been no indication of strong public concern about radioactive emissions. Radioactive discharges to the environment are subject to very strict control exercised by two ministries that do not have any administrative connection with the Nuclear Installations Inspectorate. These ministries, the Department of the Environment and the Ministry of Agriculture, Fisheries and Food, have for many years been responsible for controlling the disposal of various industrial wastes. This administrative arrangement is believed to have made an important contribution toward establishing public confidence in the safe development of electric power [ECE 1976b].

UNITED STATES

Law and Government Structure

Sites for new U.S. power plants are proposed by the utilities. Since about 90% of the utilities are nonfederal (i.e., owned by private investors, municipalities, rural cooperatives and state power projects), siting decisions are highly influenced by local and regional issues. Outside of federal water and air permits (many states have taken over this responsibility), which may require the preparation of an environmental impact statement (EIS), most regulation occurs at the state and local levels. The blend of federal, state and local regulatory involvement varies on a project basis. Table 2-4 lists examples of the environmental permits required for a modern fossil-fueled power station.

Table 2-4. Examples of U.S. Environmental Permits Required for a Modern Fossil-Fueled Electric Power Station [Corps of Engineers 1981]

	Permit/Approval[a]
Federal	
U.S. Environmental Protection Agency	National Pollutant Discharge Elimination System permit (permit for a point source water discharge)
U.S. Army Corps of Engineers	Permit for dredging of inlet channel, installing of discharge piping and construction of mooring facility
U.S. Department of Transportation, Federal Aviation Administration	Permit to construct plant and stack
State	
State Department of Environmental Conservation	Various permits covering construction and operation of circulating water system; rainwater discharge from fuel oil facility; atmospheric discharges related to the construction and operation of main and auxiliary boilers; construction of intake structure, fuel oil unloading pier and disposal pond; installation of culverts and relocation of a creek, and placement of fill material
State Public Service Commission	Approval of railroad crossings
State Commission of General Services	Approval to purchase underwater property for river construction activities
Regional Authority	
River Basin Commission	Approval of project after formal review
County	
Drainage Authority	Permit to install culverts
Highway Department	Permit for bridge construction and approval of railroad crossing
Local Communities	Building permits; approval of railroad crossing; approval to discharge treated sewage

[a]Based on Bowline Point Generating Station, Haverstraw, New York.

Regulation of Electric Utilities

In the United States and Canada, the electric power industry has formed nine regional councils to coordinate planning, construction and operation of large power supply systems. These councils form the National Electric Reliability Council. Its mission is to promote reliability and adequacy of large-scale power supplies for electric utility systems.

States regulate electric utilities by authorizing construction of generating facilities, reviewing and approving future plans, approving sites for power plants and transmission lines, ensuring reliability and adequacy of service, approving power rates, and setting rates of return on utility investments. Many state regulatory commissions consider themselves responsible for ensuring:

1. realistic electricity demand forecasts;
2. cost-effective conservation programs;
3. development of renewable energy resources;
4. protection of environmental and public health/safety interests; and
5. public participation in electric utility planning and policymaking.

On the national level, state regulatory bodies are represented by the National Association of Regulatory Utility Commissions [GAO 1981a,b].

Although the primary authority for regulating electric utilities remains with the states, several federal agencies regulate or influence various aspects of utility operations. From monitoring air quality around coal-fired generating facilities to licensing nuclear power plants, federal agencies have numerous responsibilities that have an impact on power system planning and management.

The federal Energy Regulatory Commission licenses nonfederal hydroelectric projects. It has jurisdiction over the rates charged for electricity sold on a wholesale basis in interstate commerce.

The Nuclear Regulatory Commission (NRC) regulates the construction and operation of all nuclear power plants, regardless of ownership, through a licensing process. Before licensing a new plant, NRC is required to assure there is a valid need for the power and that the proposed nuclear plant is the best alternative for meeting that need.

The Securities and Exchange Commission has jurisdiction over investor-owned electric utilities and holding companies. It controls the issuance of securities, consolidations among utilities and accumulation of assets within utilities.

The Rural Electrification Administration in the Department of Agriculture approves requests from rural electric systems for loans and loan guarantees. Thus, it assists in the construction and operation of relatively expensive rural electrical supply systems.

The Department of Energy (DOE) has promoted national energy policies and principles and develops and implements programs designed to ensure adequate and reliable supplies of energy. DOE has been responsible for assuring the reliability of electric bulk power supply and administering programs in the area of utility system planning, coordination, interconnection and rate structures. It enforces prohibitions against burning oil or natural gas in new power plants and fosters the use of coal and other alternatives to imported fuels.

The Environmental Protection Agency (EPA) establishes and enforces pollution abatement regulations to which utilities must conform. In some cases administration is at the state level.

State Regulation. Some states have enacted environmental legislation to supplement or strengthen federal law. This can compound utility problems with the permit and licensing process. Under provision of the Federal Water Pollution Control Act [(FWPCA), 33 USC 1251-1376] as amended by the Clean Water Act of 1977 (PL 95-217), the Clean Air Act (42 USC 7401, *et seq.*) and the Resource Conservation and Recovery Act [(RCRA), USC 6901-6987], all states are obligated to adopt and enforce minimum standards for protecting the quality of air, water and land use. However, states can raise their standards above the minimum federal requirements if they so desire. As a result, many environmental regulations are state-specific, and electric utilities are often confronted with different rules and regulations when they serve customers living in two or more states [GAO 1981a,b].

A key event in factoring environmental considerations and concerns into utility decision-making was the enactment of the National Environmental Policy Act (NEPA) of 1969 (42 USC 4231-4347). NEPA is regarded as the cornerstone of federal efforts in environmental protection. It requires decision-makers to take into account the probable effect that their actions (such as granting a construction permit or a power plant license) will have on the environment. From an operational perspective, the most important provision of NEPA required the preparation of an EIS for any proposed federal action significantly affecting environmental quality. EIS are required for licensing nuclear plants, hydroelectric plants and many large coal-fired plants. Each EIS must include analyses of (1) environmental impact of the proposed action; (2) alternatives to the proposed action; and (3) irreversible resource commitments that would result from implementation of the proposal [GAO 1981a,b].

Other major legislation enacted in the 1970s confirmed the federal commitment to protecting environmental quality and added new dimensions to utility planning. The FWPCA Amendments of 1972 (33 USC 1251-1376) marked a turning point in federal policy toward water pollution by ending the "right to pollute." The amendments were intended to restore and maintain the chemical, physical and biological integrity of the nation's waters. Their greatest impact on new generating plants has been in the design of cooling systems to control thermal pollution of rivers and lakes. Similarly, the Clean Air Act Amendments of 1977 (42 USC 7401-7642), which recodified federal air quality laws, established impediments to unrestricted discharge of air pollutants from electric power plants and increased industry attention to the use of pollution control equipment and "cleaner" fuels and combustion processes [GAO 1981a,b].

Air Pollution Control Strategy

Federal policy dealing with air pollution is codified in a single statute, the Clean Air Act Amendments of 1970 and 1977.

The Clean Air Act (42 USC 7401), particularly the Clean Air Act Amendments of 1970 (PL 91-604), provided the legislative authority for promulgation of regulations that have resulted in substantial costs to be borne by utility companies in controlling air emissions. Among other things, the 1970 amendments (42 USC 7409-7413) empowered EPA to establish and enforce national ambient air quality standards and to promulgate air emission standards for new and existing fossil-fuel–fired stationary sources of air pollutants. The need for this action was based on Congressional findings that a large part of the nation's pollution had spread across local and state jurisdictional lines and that growth in the amount and complexity of air pollution was resulting in mounting dangers to public health and welfare. As a result, the Congress found that air pollution prevention and control should be established "to protect and enhance the quality of the Nation's air resources so as to promote public health and welfare and the productive capacity of its population" [GAO 1981b].

EPA was required by the act [42 USC 7409(a)(1)(A)] to promulgate both primary (health protection) and secondary (welfare protection) ambient air quality standards. The Administrator was guided in setting the primary standards by 42 USC 7409(b)(1), which states:

> National primary ambient air quality standards . . . shall be ambient air quality standards the attainment and maintenance of which in the judgment of

> the Administrator, based on [air quality] criteria and allowing an adequate margin of safety, are requisite to protect the public health.

The margin-of-safety question is discussed in the act's legislative history, which indicates that the intent of the act is to protect the most susceptible group in the general population. However, the legislative history points out that the exact relationship between adverse health effects and concentrations of pollutants will be unknown [GAO 1981b; OECD 1975b,c].

EPA was also required to promulgate standards of performance (emission limitations) for new fossil-fuel–fired·stationary sources of air pollutants. These new source performance standards (NSPS) were to be developed to reduce the possibility that new sources of air emissions might contribute significantly to pollution that endangers public health and welfare.

EPA staff relied heavily on work that had been done by state agencies in setting air quality standards for particulate matter and sulfur dioxide emissions prior to 1970. The standards finally developed were based on federally prepared air quality criteria documents that outlined what was needed in setting the standards. These criteria were based on literature searches, scientific opinion and the committee agreement. The attainment of the air quality standards was the responsibility of the individual states. This was to be accomplished through state implementation plans (SIP) that indicated how each state intended to achieve the standards. Typically, each implementation plan is a compilation of state air pollution statutes, regulations and pollution control strategies that include emission limitations, land use controls and transportation controls. EPA is required either to approve the SIP (thus making them part of the federal law) or amend them in conformance with its criteria for attaining ambient air standards [GAO 1981b].

The flexibility in requiring conformance to the standards was left in large measure up to the states. Through its SIP, each state could determine which sources of pollutants would be most closely regulated. The mix was left up to the states, subject to federal approval that it appeared the national standard would be met. States could set tighter standards for new source emissions, but not looser standards than those set by EPA. If a state found, in retrospect, that the emission standards "overcontrolled," it could request review by EPA and, on approval, was allowed to relax the standards initially set. The granting of such variances is dependent on a state's ability to meet the attainment standards. In nonattainment areas, it is almost impossible for EPA to approve any variances. In cleaner areas, there is some flexibility in negotiating relaxations [GAO 1981b].

Water Pollution Control Strategy

The primary mechanism for regulating the discharge of water pollutants is the FWPCA Amendments of 1972 and 1977 (33 USC 1251). The 1972 amendments attacked a number of problems that had existed in the national strategy for controlling water pollution before 1972. This strategy based cleanup requirements on the desired uses of effluent-receiving waters as determined by individual state governments. This approach, which focused on ambient water quality standards, was generally ineffective due to a number of political, technical and legal weaknesses. The 1972 amendments were predicated on the philosophy "that no one has the right to pollute and that pollution continues because of technological limits, not because of any inherent right to use the Nation's waterways for the purpose of disposing of waste." The FWPCA established a national goal of totally eliminating pollutant discharges by 1985 and directed "that wherever attainable, an interim goal of water quality which provides for the protection and propagation of fish, shellfish, and wildlife and provides for recreation in and on the water be achieved by July 1, 1983" [GAO 1981a,b].

The definition of pollution discharges in the FWPCA included chemical waste and heat—both of which result from operating steam-electric generating plants. In addition to controlling these effluent discharges, FWPCA also specified that cooling water intake structures reflect the best technology available for minimizing adverse environmental impact.

To carry out the objectives of the FWPCA, EPA was charged with three major tasks:

1. develop and publish water quality criteria that reflect accurately the latest scientific knowledge on the kind and extent of all identifiable effects on health and welfare that may be expected from the presence of pollutants in any water body and given the criteria to the states for use in developing their water quality standards as specified in Section 303 of the FWPCA;
2. promulgate effluent guidelines so that by July 1, 1977, effluent limitations would be achieved for all specific sources of pollutants, other than publicly owned treatment plants, using the best practicable control technology and achieve further effluent limitations no later than July 1, 1987, using the best available technology with the EPA defining the technology to be used; and
3. set standards of performance for all new sources of pollution.

EPA also was given authority to issue pollutant discharge permits and prescribe conditions for such permits to ensure that provisions of the FWPCA are carried out. FWPCA also provides that such authority can be exercised by individual states subject to review and approval by the EPA [GAO 1981b].

Regulations for controlling thermal discharges were a combination of federally developed water quality criteria and state-developed water quality standards, which were reviewed and revised according to changes in the water quality criteria. The first criteria were issued in 1972 and were simply an updated version of those published in 1968 by the Department of Interior. EPA recognized that water quality criteria are changeable and that criteria development is a continuing, progressive research effort. Consequently, a further update was issued in 1976. EPA also had the responsibility for developing regulations that established effluent limitation guidelines for existing utilities and standards of performance for new steam-electric power generating units. These guidelines were established within the framework of the technology available to the utility industry to achieve compliance [GAO 1981b].

Noise Control Strategy

The Noise Control Act of 1972 directs EPA to promote an environment for all Americans free from noise that jeopardizes their health and welfare. It specifies that EPA regulate new products in commerce and establish noise labeling requirements for noisy products and for products designed to reduce noise. The act encourages the development of noise control programs on the community and state level. To date EPA has focused regulatory efforts on the control of transportation and construction noises.

The Occupational Safety and Health Administration (OSHA) is charged with developing and enforcing rules to prevent excessive noise exposure. OSHA's mandate is that no employee should suffer material impairment of health or functional capacity. Under OSHA's purview is a federal noise standard to protect workers from noise exposure. The maximum limits are a function of exposure time.

Before the 1970s, there was no special provision for noise associated with power plants. In response to the development of noise standards in the 1970s, noise control in power plants became a matter of attenuation. In the 1980s, the industry is looking ahead to the need to comply with specific equipment standards for the control of noise [EPRI 1980a].

Solid Waste Control Strategy

Before 1976, the main federal laws relating to solid waste disposal were the Solid Waste Disposal Act (which pertained mostly to municipal wastes) and the federal Pesticide Control Act and Amendments (which could be used to regulate bactericides and fungicides in power plant cooling waters). In 1976 two acts pertinent to power plant wastes were promulgated: RCRA and the Toxic Substances Control Act (TSCA).

Solid waste disposal at the end of the 1960s was accomplished primarily by ponding or trucking the waste to state-licensed sanitary landfills. Today, two classes of solid waste are recognized by the RCRA: hazardous and nonhazardous. A procedure for distinguishing between these two categories is provided together with different regulatory requirements. The hazardous criteria for solid wastes (as well as air and water pollutants) are being guided by the definition of toxic substances under TSCA. New disposal requirements may be mandated even if the waste is declared nonhazardous because of heightened concern for groundwater contamination and human health impacts of toxic substances. The combined impact of these acts may increase the cost of fly ash and scrubber sludge disposal to a level that eliminates the use of sludge-producing scrubbers and forces the use of regenerable desulfurization systems [EPRI 1980a].

The May 19, 1980, hazardous waste regulations exempted high-volume utility wastes from the more stringent hazardous waste manifesting and record-keeping requirements for a period of three years. The exemption included fly ash, bottom ash, slag wastes and sludges from FGD. During the exemption period, EPA is conducting a comprehensive study of utility waste disposal practices. In the interim, the industry still must comply with the nonhazardous waste disposal regulations promulgated by EPA and the states [EPRI 1980a].

FGD systems will require the construction of waste disposal ponds or landfills for sludge disposal. Use of wet disposal systems usually will require liners under the ponds to protect the underlying aquifers from leachate migration. Any overflow from these disposal ponds must be collected and treated before discharge. Liners and leachate recovery systems in landfills are necessary at some sites, depending on the permeability of the native soil [EPRI 1980a].

Power Plant Siting Strategy

The main tool used in siting decisions is the EIS. Required in compliance with the 1969 NEPA, it is prepared for all major federal actions significantly affecting the environment. Such actions with respect to power plants have included licensing by the NRC and permitting by the EPA and the U.S. Army Corps of Engineers.

Of particular importance to siting is the fact that the EIS must not only review significant adverse environmental impacts of the proposed siting, but also must identify and comment on alternatives to the project as proposed. This includes the option of not constructing the proposed plant. The EIS can result in project modifications or even rejection of the project on environmental grounds.

The EIS process acts as a mechanism of public education regarding a proposed facility. It also provides a mechanism for public participation and comment by public hearings and written review comments on draft versions of the EIS.

Some states have passed legislation that allows a site analysis hearing to be held and a permit to construct to be issued simultaneously with all other state permits. However, the applicant's EIS often must be circulated individually to every relevant agency, resulting in separate issue of permit approvals.

About 20 state utility commissions grant licenses, permits or certificates for new thermal power plants. All of these consider the principal concerns in the project review to be reliability and safety. Half of the commissions that grant licenses consider visual amenity and the appearance or location of transmission lines. Other issues, such as air pollution, air navigation, fish and wildlife, recreation, and water pollution (thermal) are the concern of other state agencies. In states where the utility commissions do not license power plants, other agencies account for the full range of siting factors [OECD 1977].

The Federal Power Act provides for the licensing of hydroelectric plants. Within its purview, the federal Energy Regulatory Commission (formerly the Federal Power Commission) has sought views from the Department of Interior and has focused considerable attention on water pollution control requirements in hydroelectric project licenses [OECD 1977].

In the United States, the role of the national government is, as in other countries, much more influential where nuclear power stations are concerned. In the United States, construction permits and operating licenses for nuclear power plants must be obtained from the NAC.

The power reactor licensing procedures that precede issuance of a construction permit consist of four steps. First, there is an informal site evaluation, whereby the prospective applicant discusses with the NRC the suitability of various reactor sites under consideration. Second, the application for a construction permit is prepared by the utility (usually with the help of the reactor supplier) and submitted to the NRC. This application includes the preliminary design and safety features of the proposed reactor, and comprehensive data on the proposed site. This application is then reviewed by the NRC regulatory staff and by the Advisory Committee on Reactor Safeguards. The third step is a public hearing, conducted in the vicinity of the project by an atomic safety and licensing board. An initial decision on the suitability of the proposed site for the planned use is tendered. This is subject to review by the NRC, either on its own initiative or on petition by a party to the proceedings. The review process occurs again before the operating license is issued; however, a public hearing usually is not held [OECD 1977].

Public Apprehension to the Siting of Power Plants

In the 1960s the public utility industry was characterized by sales and revenues growing at steady rates, and construction focused on economies of scale through the construction of large power stations. This resulted in reduced consumer prices, and environmental or social impacts were of concern only to a few people. The regulatory process faced by the industry was relatively simple, and controversies over electric power plans and policies were rare [GAO 1980a].

The 1970s brought about substantial change. The large size, costs, and environmental and safety impacts of new generating facilities focused considerable attention on the industry. Most new power plants were so large and costly that their siting and funding became regional issues. The 1969 NEPA, requiring preparation of an EIS, became a central point of reference for the expression of public concerns. The EIS process played a critical role in creating, defining and clarifying channels for public participation. Still, public opposition increased markedly, especially in the area of citizen suits (an action unique to the United States with respect to electric power). Within the first six years of the decade, public action through the courts obtained the stoppage of work on 20 stations in 12 states [Commission of the European Committees 1979; GAO 1980a; OECD 1977,1980].

Concerns have substantially increased as a result of the relatively recent accident at the Three Mile Island nuclear plant in Pennsylvania. As a result of this accident and various other incidents and problems at nuclear power stations, both existing and under construction, negative public attitudes present a substantial barrier to the timely siting and construction of new nuclear power stations.

CHAPTER 3

POLLUTION CONTROL COSTS

The most generally recognized sources of increase, in real terms, in the cost of electricity production are those resulting from conformance with air, water, solid waste and other pollution control regulations. Low-sulfur fuels, stack gas cleaning equipment, and cooling towers or ponds are expected to cost producers more and result in higher prices over those encountered in the absence of relevant environmental protection statutes.

Other environmental considerations, such as esthetics, the preservation of wildlife and long-term global air pollution problems are more difficult to reflect as costs, but the costs of improving esthetics are being reported. Other measures, such as underground transmission lines in the vicinity of a power station and modified siting practices, also can cause significant cost increases.

A considerable amount of the literature is devoted to the advantages of pollution control, either from savings as a result of reduction in damage to health, crops and building materials or from an increase in the employment rate. For example, the Federal Republic of Germany has estimated that the installation of flue gas desulfurization (FGD) equipment will provide 24,000 man-years of work and its operation will require an additional 500 permanent workers. According to a study of European countries, as much will be saved by reduced damage as will be spent on reducing sulfur emissions. Thus, high expenditures for pollution control are justified from an economic point of view [OECD 1981; Scharer 1980].

Although many studies suggest a relationship between total expenditures for pollution control and personnel requirements, the relationship is difficult to determine accurately. This is because of the varying nature of pollution control activities from industry to industry and from medium to medium. For example, water pollution control activities appear to be more labor-intensive than do air pollution activities. Also, thrusts of pollution control research and development (R&D) change rapidly, as do the financial commitments necessary to maintain a man-year of research effort [NAS 1977].

Certain realities confront a utility when it constructs nonrevenue-producing pollution abatement facilities. "Nonrevenue-production" refers to equipment that does not increase the utility's installed capacity and therefore its ability to expand service. To the extent that the cost of the pollution control facility is included in the utility's rate base, it is or becomes revenue-producing as soon as further rate relief is obtained. Ordinarily, if a utility can earn more than its capital cost, the incentive would be to add the capital equipment, particularly if a rate commission would permit the utility to earn interest during construction [Berlin et al. 1974].

When such capital expenditures are to be recouped within one year, they usually are treated as operating expenses. When recoupment is longer and when R&D is undertaken by the utility itself, usually the utility is allowed to treat the expenditures as construction work in progress. In this case it may be permitted to earn interest during construction, but at a rate below the authorized return [Berlin et al. 1974].

When R&D funds are contributed to projects undertaken by others (e.g., the Electric Power Research Institute), the utility would be permitted to amortize the expense over a five-year period and such costs could be treated as operating expense and no interest would be realized [Berlin et al. 1974].

It is easy to address only the impact of pollution controls on plant capital costs, but improved controls also affect fuel costs, operating and maintenance (O&M) costs, and performance reliability (capacity factor). The heat, steam and electricity required to run pollution control equipment reduce thermal efficiency and increase fuel consumption. O&M costs are raised by the limestone and other material requirements of scrubbers, disposal costs for ash and sludge, and the personnel needed to operate control devices. Also, breakdowns in control equipment or gas and moisture carryover can impair plant availability, although increased use of redundant scrubber modules is reducing this effect [Komanoff 1980].

COSTS OF ALTERNATIVE POLLUTION CONTROLS

There are a number of alternatives available for controlling air pollution from power stations. When the costs of these alternatives are examined, there are wide ranges in capital, operating and annual costs for different technologies. For example, physical coal cleaning (a control widely used in the United States compared to European countries) may be less than one-sixth as costly as flue gas cleaning with respect to capital investment dedicated to sulfur removal. However, the technique is only partially effective and is not useful on all types of coals.

Table 3-1 presents a comparison of the costs of various air pollution control techniques. Development in control technology will significantly affect these cost estimates as the technologies are refined or replaced with less costly alternatives.

Extrapolation of costs of pollution control for a nation based on case or unit process examples is difficult. This is because substantial economies of scale are attained as boiler size increases. Also, wide variations occur among various pollution control systems, especially at lower boiler capacities (Figure 3-1). For this reason, estimates and inferences of pollution control costs for an entire country based on case examples are not given in this chapter. The information presented is based on reported historical data or on estimates of overall pollution control costs by officials of the countries addressed.

ESTHETIC CONSIDERATIONS

A trend that began at the beginning of the last decade and is continuing is public opposition to the siting of large power stations. In the United Kingdom, such public opposition has led to a serious impediment to the siting of new large power stations in rural areas of England. Of the countries addressed in this report, public opposition to the construction of power stations based on esthetic grounds is most pronounced in the United Kingdom and least in the United States.

The main facilities that cause apprehension are the requirements for land, large cooling towers, and high-voltage transmission lines and their supporting towers. Assuming onsite coal and ash storage or fuel storage as appropriate, the land requirements for a 3000-MWe generating station can vary from 1200 ac for a station powered by coal to 400, 350 and 200 ac for stations powered by nuclear fuel, oil and gas, respectively [Scott 1973].

Because of public opposition based on esthetic grounds, a considerable amount of the costs of construction of new facilities is attributable to esthetic considerations. In the United States, almost one-fourth of the in-service capital costs applicable to environmental protection facilities are associated with abating negative esthetic impacts [DOE 1981a].

Table 3-1. Comparison of Alternative Air Pollution Controls [EPA 1978]

Control Technology and Removal Efficiency (%)	Capital Equipment ($/kW)	Operational Costs (mill/kWh)	Total Annualized Cost (mill/kWh)	Pollution Control Costs as a Percentage of Capital Investment
For Control of Sulfur Dioxide				
Flue Gas Desulfurization (85%)	60–85[a]	2.1–3.6	3.4–5.4	13–19[b]
Physical Coal Cleaning (20–40%)	9–22	0.15–1.20	1.5–2.0	2–4
For Control of Nitrogen Oxides				
Combustion Modification (20–60%)	0.5–7[c]	0.01–0.35	0.005–0.030	0.1–2
For Control of Particulate Matter				
Electrostatic Precipitators (98%)	30–90	0.04–0.07	0.9–2.8	7–20
Fabric Filters (99%)	38–48	0.01–3.0	1.5–2.5	8–10
Wet Scrubbers (80–98%)	49	0.4	2.0	11

[a]Costs based on installation in new units, 1977 dollars.
[b]Based on total cost of $450 million for a 1000 MWe plant.
[c]This cost based on 1975 dollars.

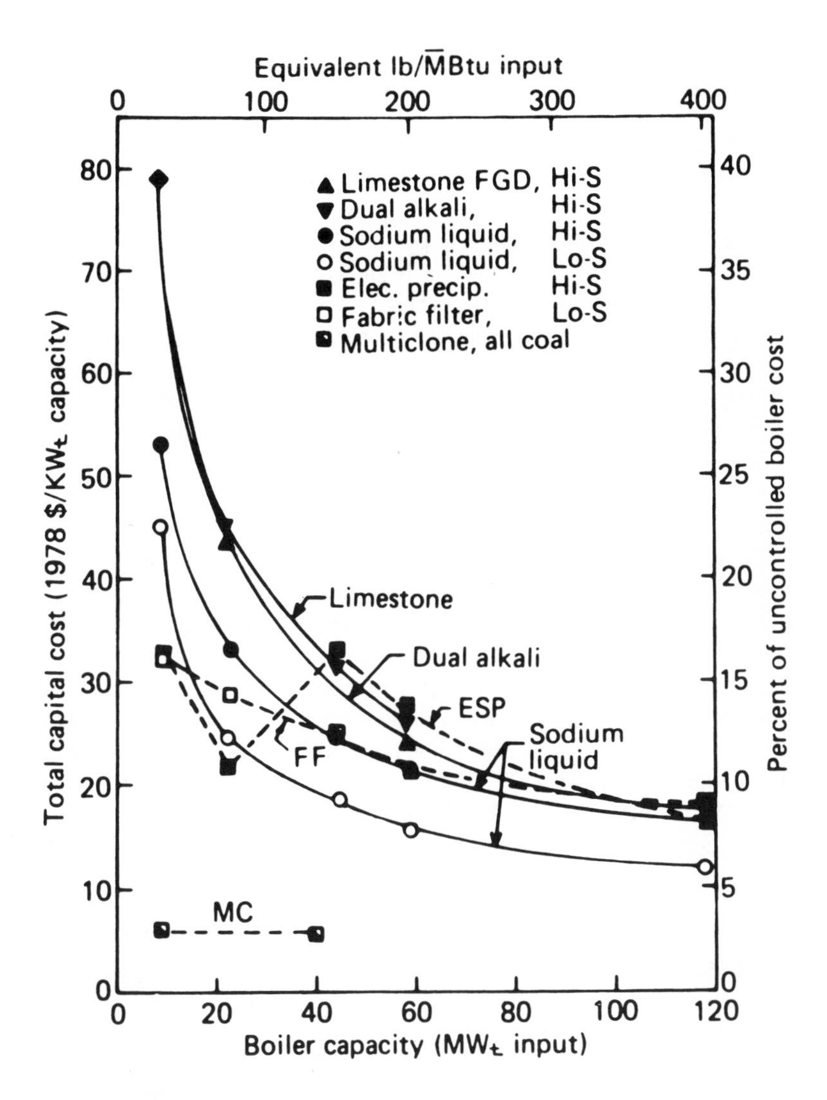

Figure 3-1. Cost of alternative air pollution control systems. Data represent normalized capital costs for industrial boilers [Rubin 1981].

Since the issue of esthetic impacts is considered a major environmental problem in the countries addressed in this book, it is addressed separately. Appendix A discusses public opposition and corrective measures relative to esthetics that are being taken within the countries.

VARIATIONS IN COSTS RELATED TO TYPE OF FUEL

Costs for pollution abatement can vary among power plants of the same capacity, but powered by different fuels. The differences are shown in Table 3-2.

Both fossil-fueled and nuclear power plants require thermal pollution control. About 10% of the waste heat from fossil-fueled plants is discharged through the stacks, whereas nearly all heat is released from nuclear plants through cooling water. As a result, nuclear plants release nearly 50% more waste heat into water than comparable sized fossil-fueled plants and thus incur substantially more costs for thermal control.

The need for air pollution control rather than radiation control is associated with fossil-fueled power plants; the opposite relationship exists with respect to nuclear plants. The need for esthetic improvements is associated with both types of power stations.

The data presented in Table 3-3 are a rough estimate of the relative costs of controlling pollution in the electric power industry. Exact data are not useful as overall indicators, because control devices vary markedly on a site-specific basis. For example, Table 3-3 lists the percentages of the component costs of different types of 1000-MWe steam power plants constructed in 1980 on a hypothetical site. These are based on 1973 estimated costs of these plants to be completed in 1980. The hypothetical total costs were \$435/kW for a nuclear plant, \$409/kW for coal, \$377/kW for oil and \$285/kW for gas. Depending on the location in the United States, the estimated capital costs for the nuclear plant varied

Table 3-2. Effect of Pollution Abatement on the Total Costs of New Fossil-Fuel vs Nuclear Power Plants[a] [Scott 1973]

Type of Control	Type of Plant	
	Fossil Fuel	Nuclear
Thermal Control (Wet Towers)	1.8	2.7
Air Pollution Control		
Particulate Matter	1.0–2.0	
Sulfur Oxides	3.2–6.5	
Radioactivity Control		1.0
Esthetic Improvements	3.2	3.2
Total	9.2–13.5	6.9

[a] Values are in terms of percent increase in new plant construction costs attributable to a particular pollution abatement activity. They are rough estimates of the proportion of typical costs. Relative percentages associated with a specific site can be higher or lower depending on local regulations, water availability and other factors.

Table 3-3. Percentages of Component Costs of Different Types of Steam-Electric Power Plants[a] [Olds 1973]

Component	Nuclear	Coal	Oil	Gas
Land	0.2	0.2	0.3	0.4
Structures and Site Facilities	8.7	6.1	6.1	8.1
Reactor/Boiler Plant Equipment	13.6	15.2	13.5	11.6
Turbine Plant Equipment	15.6	12.7	13.8	18.2
Electric Plant Equipment	3.7	3.4	3.2	4.2
Miscellaneous Plant Equipment	1.1	1.0	1.1	1.4
Contingency Allowance	2.8	2.7	2.6	2.8
Spare Parts Allowance	0.2	0.2	0.3	0.4
Indirect Costs	11.3	8.6	8.5	10.2
Escalation to Start Construction[b]		4.9	4.8	5.6
Escalation During Construction[b]	22.8	16.9	16.7	19.2
Interest During Construction	14.5	11.0	10.9	12.6
Cooling Towers	4.6	3.9	3.9	5.3
Near-Zero Radwaste System	0.9			
Sulfur Dioxide Removal System		13.2	14.3	
Total	100.0	100.0	100.0	100.0
Estimated Total Costs ($/kW)	435	409	377	285

[a] All data refer to capital estimates for a hypothetical 1000-MW steam-electric power station to begin operation in 1980 with construction beginning in early 1971 for a nuclear plant or 1974 for a fossil-fuel plant.

[b] Escalation rates are 5%/yr for equipment and materials and 10%/yr for labor.

$376–465/kW; coal, $349–440/kW; oil, $319–405/kW; and gas $240–306/kW [Olds 1973].

The main components of power plant costs (i.e., capital, O&M, and fuel costs) vary in their relative magnitude when nuclear- and fossil-fuel–fired plants are compared. For example, the capital cost for a nuclear power plant is much higher in proportion to total costs than for a coal-fired power plant. A coal-fired power plant requires a relatively greater percentage of costs dedicated to O&M, partly due to air pollution control equipment operation and monitoring (Table 3-4).

A comparison of the annual operating costs of a typical 1000-MWe power plant using different types of fuel is presented in Table 3-5. The operating costs associated with the protection of human health and the environment as a percent of conventional costs is highest in a coal-fired power plant, about 33% less in either an oil-fired or nuclear facility, and the lowest in a gas-fired plant. The proportion of abatement costs are more striking.

The International Institute for Applied Systems has prepared a non-quantitative estimate of the overall relative capital and operating costs for all of the major types of electrical generating facilities. The estimates of

Table 3-4. Relative Differences Among Nuclear and Coal-Fired Power Plant Subcosts

	Power Source	
	Nuclear	Coal-Fired
Capital Investment (%)	66.42	48.68
Operation and Maintenance (%)	26.68	44.70
Fuel (%)	6.90	6.62
Total (%)	100	100

relative capital and operating costs are given in Tables 3-6 and 3-7, respectively.

With respect to nuclear power, it appears to be difficult to determine the additional cost per kilowatt of installed power that can be attributed to environmental requirements. In a 1974 survey conducted by the Group of Experts on Electric Power and the Environment, of the Economic Commission for Europe, only three replies were received with respect to costs associated with environmental project requirements. The answers received place environmental protection costs at 0.1% of capital investment and 0.5% of operating costs. The survey included the Federal Republic of Germany, France and the United Kingdom, of which only France replied with information on environmental costs, and they were associated with monitoring. The replies relating to environmental costs are provided in Table 3-8 [ECE 1977a].

Recently, France has compiled an estimate of operating costs for pollution control equipment at hydroelectric, thermal and nuclear facilities. France also has classified the costs by control category for each type of power plant for the period 1970–1980. These costs are presented in the section that deals with pollution control in France.

FEDERAL REPUBLIC OF GERMANY

Overall Costs

The Federal Republic of Germany embarked on a major program of pollution control during the 1970s. The total expenditure for all economic sectors during 1971–1975 to comply with the program was projected to be about 36.0 billion deutschmarks, or about 1% of the gross national product (GNP). The costs of capital investments of the program to control pollution from power generation during the period were estimated to be 500 million deutschmarks, or about 2% of the 28 million deutschmarks in capital investments in the program for the period [OECD 1972].

Table 3-5. Comparison of Estimated Costs[a] of Alternative Energy Systems[b]

	Coal	Oil	Gas	LWR[c]
Conventional Costs				
Capital Plant	51	47	42	77
Fuel Cycle	64	180	237	39
Operation and Maintenance	5.4	4.0	3.7	5.2
Rounded Totals	120	231	283	121
Abatement Costs				
Cooling Towers	3.6	5.9	7.1	2.4
Sulfur/SO_2 Removal	25.9	4.9	NA[d]	NA
Strip-mined Land Reclamation	0.1	NA	NA	S[e]
Near-zero Radwaste	NA	NA	NA	1.2–1.8
Rounded Totals	30	11	7	3–4
Conventional and Abatement Totals	150	242	290	125
Abatement Component (%)	20	5	2	3
Safety				
Occupational[f]	0.46	0.086	0.039	0.05
Public[g]	0.18	U[h]	U	0.003
Subtotal	0.64	0.086	0.039	0.053
Health				
Occupational	0.03[i]	U	U	0.024[j]
Public	U	U	U	0.01[k]
Subtotal	0.03	U	U	0.034
Total Human Health and Accident Costs	0.067	0.086	0.039	0.087
Environmental Effects				
Water Base	0.4	0.4	0.4	0.6
Air Base	0.8	0.6	0.1	S
Land Base	0.2	S	S	S
Subtotal	1.4	1.0	0.5	0.6
Total Human and Environment Effects	2.1	1.1	0.5	0.7
Percent of Conventional Costs (%)	3	1	0.5	0.9

[a] Millions of 1980 dollars/year, unless otherwise noted.
[b] Annual operation of 1000-MWe power plant and supporting fuel cycle (6.57×10^9 kWhe).
[c] LWR = light-water reactor.
[d] NA = not applicable.
[e] S = small.
[f] Conventional injuries in routine industrial accidents, including fatal and nonfatal injuries: 1 death = 6000 man-days lost = $300,000.
[g] Coventional injuries in accidents in transportation of fuels: 1 death = 6000 man-days lost = $300,000.
[h] U = unevaluated.
[i] Coalworker's pneumoconiosis.
[j] Radiological health effects, including lung cancers among uranium miners.
[k] Radiological health effects from routine emissions.

Table 3-6. Relative Capital Costs for Unit of Electricity Generation by Energy Source [International Institute for Applied Systems Analysis 1981]

Energy Source	Relative Power Plant Cost per Unit of Output[a]
Natural Gas	Very low
Oil	Low
Coal	Moderate
Nuclear	
Light Water Reactor	Moderate, but greater than coal
Heavy Water Reactor	Moderately high
Hydropower	High

[a] These reflect costs involved with processing fuels or with converting fuel to electricity, but do not include the fuel itself. They are only for electrical generation. They include labor and material costs in mining, transportation and fuel preparation. For nuclear, they include the operating costs of fuel purification, enrichment, fabrication, reprocessing in certain fuel cycles and waste disposal. For coal, they include transportation, boiler feed preparation, stack gas cleaning and waste disposal.

Table 3-7. Relative Operating Costs for Unit of Electricity Generation by Energy Source [International Institute for Applied Systems Analysis 1981]

Energy Source	Relative cost Per Unit of Generation[a]
Hydropower	Very low
Natural Gas	Low
Oil	Low
Coal (near mine)	Moderately low
Nuclear	Moderate
Coal (far from mine)	High

[a] These reflects costs involved with processing fuels or with converting fuel to electricity, but do not include the fuel itself. They are only for electrical generation. They include labor and material costs in mining, transportation and fuel preparation. For nuclear, they include the operating costs of fuel purification, enrichment, fabrication, reprocessing in certain fuel cycles and waste disposal. For coal, they include transportation, boiler feed preparation, stack gas cleaning and waste disposal.

In 1975 about 247 million deutschmarks in capital investments were made by the German public utility industry to control pollution. This amount represented 1.8% of total capital investments of the utilities (electricity, gas, water supply and district heating). Of this amount, 0.82% was spent on controlling air pollution, 0.46% on water pollution, 0.14% on solid waste management and 0.37% on noise control [Statistisches Bundesamt undated].

Table 3-8. Environmental Costs Associated with Nuclear Power Stations [ECE 1977a]

Country	Replies[a]
Czechoslovakia	0.1% of total capital investment (0.96 $/kW installed)
France	0.5% of the operating costs
Spain	Capital costs: boiling water reactor, 0.99 $/kW installed; pressurized water reactor, 0.58 $/kW installed Operating costs: boiling water reactor, 0.14×10^{-4} $/kWh; pressurized water reactor, 0.96×10^{-5} $/kWh

[a] Results based on replies to a 1974 survey conducted by the Economic Commission for Europe. The environmental costs cited are mostly attributable to the cost of treating thermal effluents (capital investment) and of monitoring liquid discharges (operating costs).

Table 3-9. Costs of Air Pollution Control Equipment Installed in an Electric Power Station in the Federal Republic of Germany [UmweltsBundesAmt 1980a]

Category	Cost[a] (millions of deutschmarks)
Capital Investment	115.0
Operation Costs	
Depreciation, interests, taxes, insurance, administration, maintenance and repair	23.0
Labor	0.8
Materials, electricity	13.6
Total annual costs	37.4
Costs per kWh	0.01

[a] Calculation based on costs of flue gas desulfurization as of 1978 of a coal-fired power plant with a capacity of 700 MWe (calorific capacity of 7.8 TJ/hr) attaining 80–95% sulfur removal and operating at an average of 5200 hr/yr.

Costs of Air Pollution Control

To determine the costs of controlling sulfur dioxide air pollution from power plants, the Federal Environmental Agency recently conducted a survey of German manufacturers of FGD equipment with respect to relative efficiency and cost data. The calculations were based on the characteristics of a modern, coal-fired 700-MWe power plant. The results are summarized in Table 3-9.

According to the results of the survey, the total expenditure of about one billion deutschmarks required for the construction of a coal-fired 700-MWe power plant includes about 115 million deutschmarks for FGD. The annual desulfurization cost (including service of capital) amounts to 37 million deutschmarks. These costs would increase the price of the electricity leaving the plant by about 0.01 deutschmarks/kWh. However,since up to 1990 only about 10–20% of the power supply will come from desulfurized power plants, the final price increase to be expected by the consumer will not exceed 10–20% of the above amount [UmweltBundesAmt 1980].

In current deutschmarks, about 37 million marks are required for the annual operation and maintenance of flue gas scrubbers at a new 700-MWe coal-fired electric power station. Of this amount, 23 million marks represents depreciation, interest and insurance on the original capital investment (average 115–120 million marks). These costs reflect compliance with proposed air pollution control regulations expected to be promulgated in May 1982 [UmweltBundesAmt 1982].

FRANCE

In France, investment in air or water pollution facilities and equipment qualifies for accelerated depreciation. In addition, the Ministry for Industrial and Scientific Development may aid industry by providing 50% of the costs to develop new products or processes destined for pollution abatement. However, the grant must be reimbursed if the product or process becomes commercially feasible [Congressional Research Service 1981].

Of the total expenses for environmental protection associated with the generation of electricity in France over the period 1970–1980, about 40%, or 1.966 billion tax-free 1980 francs, was spent on capital investments, and about 41% (1.987 billion francs) was spent on O&M. About 19% was spent on R&D. The percentage of power station environmental control costs of the total pollution control for production, transmission and distribution rose from an average of 77.4% in 1970 to a peak of 90.6% in 1976 and diminished to about 84% in 1980 (Table 3-10).

With respect to capital investments in pollution control equipment and facilities for electric power stations during the period 1970–1980, about 18% was spent on air pollution control, 30% on water pollution, 17% on noise control, 3% on esthetics and 32% on control of radiation. The percent of capital pollution control costs associated with electricity generation out of the total investments for electric utilities (including

Table 3-10. Total Expenses for Environmental Protection Associated with Electric Power Stations in France[a] (1970–1980) [EDF 1982]

Expenses	1970	1971	1972	1973	1974	1975	1976	1977	1978	1979	1980
Capital Investments	46.2	58.4	81.8	119.4	108.7	178.8	171.2	213.7	275.9	321.7	389.6
Operating Costs	91.5	86.4	85.4	122.5	226.1	160.2	289.5	138.7	191.0	287.0	309.0
Research & Development Costs	12.6	16.4	30.7	34.3	57.8	70.7	132.7	129.5	137.2	126.7	158.2
Total	150.3	161.2	197.9	276.2	392.6	409.7	593.4	481.9	604.1	735.4	856.8
Total Environmental Expenses (including transmission and distribution)	194.1	211.7	237.3	351.1	447.2	463.3	654.8	549.2	731.8	863.2	1,071.2
Percent of Total Environmental Expenses Attributable to Power Stations	77.4	76.2	83.4	78.7	87.8	88.5	90.6	87.8	83.6	85.2	84.2

[a] All costs expressed as millions of 1980 tax-free francs.

Table 3-11. Capital Investments in Pollution Control Equipment and Facilities for Electric Power Stations in France[a] (1970–1980) [EDF 1982]

Pollution Control Category	1970	1971	1972	1973	1974	1975	1976	1977	1978	1979	1980
Air	18.9	27.3	40.7	65.3	39.1	40.5	19.6	16.0	14.1	7.2	56.5
Water	9.6	15.6	15.2	17.4	22.5	15.9	48.1	71.1	116.2	101.4	157.3
Noise	9.1	7.2	9.2	13.4	13.5	89.1	31.5	28.9	28.7	37.0	61.2
Esthetics	4.0	4.2	8.7	3.8	3.1	5.6	2.3	13.6	7.5	6.1	5.4
Radiation	4.6	4.1	8.0	19.5	30.5	27.7	69.7	84.1	109.4	170.0	109.2
Total	46.2	58.4	81.8	119.4	108.7	178.8	171.2	213.7	275.9	321.7	389.6
Power Station Pollution Control Costs as a Percent of Total Capital Investments	0.6	0.7	0.9	1.3	1.1	1.5	1.3	1.5	1.5	2.6	2.6
Transmission Pollution Control Costs as a Percent of Total Capital Investments	0.5	0.6	0.4	0.8	0.5	0.4	0.4	0.5	0.7	1.0	1.1
Total Capital Investments (millions of 1980 francs)	8,182	8,377	9,323	9,252	10,206	12,232	13,606	13,940	18,223	12,416	14,800

[a] Investment costs are in millions of 1980 tax-free francs. Percentages are of the total investments for electric utilities. Total capital investments include generation and transmission costs.

transmission) rose from 0.6% in 1970 to 2.6% in 1980. Pollution control costs of electric utilities not associated with electricity generation only rose from 0.5% in 1970 to 1.1% in 1980 (Table 3-11).

With respect to pollution control operating costs for power stations during the period 1970–1980, about 46% was associated with air pollution control and 52% with water pollution control. Only 1% of the pollution control operating costs were associated with radiation control and 1% with improving esthetics (Tables 3-12 to 3-14).

Of all of the pollution control operating costs for power stations during the period, 44% was dedicated to hydroelectric power, 51% to thermal (fossil-fuel) power and 5% to nuclear power. Most of water pollution control costs were associated with hydroelectric power, whereas most air pollution costs were associated with thermal (fossil-fuel) power. Expenditures associated with radiation management were strictly dedicated to nuclear power (Table 3-14).

Table 3-12. Percentage of Pollution Control Operating Costs for Power Stations in France Broken Down by Pollution Control Categories and by Energy Source (1970–1980) [EDF 1982]

Categories	Total Operating Costs (Millions of 1980 Francs)	Percent
Pollution Control		
Air	914.1	46
Water	1043.0	52
Radiation	17.0	1
Esthetics	10.6	1
Noise	2.6	
Total	1987.3	100
Energy Source		
Hydro		
Water	848.5	43
Esthetics	10.6	1
Total	859.1	44
Fossil		
Air	914.1	46
Water	107.8	5
Noise	2.0	
Total	1023.9	51
Nuclear		
Water	86.7	4
Radiation	17.0	1
Noise	0.6	
Total	104.3	5
Total	1987.3	100

Table 3-13. Operating Costs of Pollution Control Activities for Electric Power Stations in France (1970–1980) Divided into Pollution Control Categories[a] [EDF 1982]

Pollution Control Category	1970	1971	1972	1973	1974	1975	1976	1977	1978	1979	1980
Air	44.5	34.1	36.7	70.9	121.5	72.7	81.6	71.2	82.2	156.5	142.2
Water	45.3	50.4	46.2	49.1	102.4	85.0	206.4	65.7	105.8	124.6	162.1
Noise			0.5	0.5		0.2	0.2	0.4	0.2	0.2	0.4
Radiation	1.4	1.6	1.7	1.7	1.9	2.0	0.9	1.0	1.1	2.1	1.6
Esthetics	0.3	0.3	0.3	0.3	0.3	0.3	0.4	0.4	1.7	3.6	2.7
Total	91.5	86.4	85.4	122.5	226.1	160.2	289.5	138.7	191.0	287.0	309.0

[a]Costs are in millions of 1980 tax-free francs.

Table 3-14. Operating Costs of Pollution Control Activities for Electric Power Stations in France (1970–1980) Divided into Energy Source and Pollution Control Categories[a] [EDF 1982]

Pollution Control Category	1970	1971	1972	1973	1974	1975	1976	1977	1978	1979	1980
Hydropower											
Water	44.3	49.3	45.0	46.1	81.3	72.2	56.5	65.3	105.6	122.0	160.9
Esthetics	0.3	0.3	0.3	0.3	0.3	0.3	0.4	0.4	1.7	3.6	2.7
Total	44.6	49.6	45.3	46.4	81.6	72.5	56.9	65.7	107.3	125.6	163.6
Fossil											
Air	44.5	34.1	36.7	70.9	121.5	72.7	81.6	71.2	82.2	156.5	142.2
Water	1.0	1.1	1.2	3.0	3.0	3.8	90.4	0.3	0.2	2.6	1.2
Noise			0.5	0.5		0.2	0.2	0.2	0.1	0.1	0.2
Total	45.5	35.2	38.4	74.4	124.5	76.7	172.2	71.7	82.5	159.2	143.6
Nuclear											
Water					18.1	9.0	59.5	0.1			
Radioactivity	1.4	1.6	1.7	1.7	1.9	2.0	0.9	1.0	1.1	2.1	1.6
Noise								0.2	0.1	0.1	0.2
Total	1.4	1.6	1.7	1.7	20.0	11.0	60.4	1.3	1.2	2.2	1.8
Total Operating Costs	91.5	86.4	85.4	122.5	226.1	160.2	289.5	138.7	191.0	287.0	309.0

[a]Costs are in millions of 1980 tax-free francs.

UNITED KINGDOM

With respect to the United Kingdom, costs of pollution control in terms of reallocation of capital resources generally fall on the polluter. Britain subscribes to the polluter-pays principle. However, nationalized industries receive long-term financing to help capital-intensive programs, as well as additional assistance for operating costs. Pollution control equipment is eligible for tax exemptions. In "development areas," government grants may be available for new, privately owned facilities [Congressional Research Service 1981].

In the United Kingdom, tall chimneys are used as a pollution control strategy to decrease ground-level concentrations of air pollutants. This policy has resulted in a relatively low capital investment. The tall chimney and the 99.3% efficient precipitator contribute less than 3% to the total capital cost of a modern coal-fired plant. This investment may be reasonable in view of the high air quality achieved without other apparent environmental problems beyond the appearance of the chimney and the disposal of ash. In contrast, the removal of sulfur dioxide, which would still necessitate a tall chimney, would increase the percentage of the capital costs devoted to environmental protection by an order of magnitude (30%), require large quarrying activities in Britain's limestone hills, and would pose a severe "solid" waste disposal problem [Leason 1974].

The discharge of warmed cooling water is not considered a major problem. If, however, Regional Water Authorities cannot increase water resources in pace with new generating station requirements, and/or the charges for providing water to be evaporated become prohibitive, inland stations may need to be equipped with dry towers, which are about four times as expensive as current natural-draft wet towers. A lowering of permitted discharge temperature to protect a sewage-polluted river would not only increase generating costs, but would deviate from conventional policy [Leason 1974].

It is possible to lower the sulfur content of fuel oil by established catalytic desulfurization processes, but the cost is high, especially for the residual fuels largely used in power stations. Catalyst life is short with the ash-bearing, high-carbon-content residuals. The alternative of further concentrating the residue, desulfurizing the volatile extract and reblending, seems to be gaining acceptance. Whichever method is adopted in England, the increase in costs is unlikely to be less than 30% of the original price for reducing the oil from 4 to 1% sulfur. Removed from the oil, the sulfur can be recovered in its elemental form, which minimizes the problems of disposal [Leason 1974].

The cost of pollution control equipment in coal-fired electric plants in the United Kingdom is about 5–10% of plant capital costs. Component cost breakdowns for a typical 2000-MWe coal-fired power plant are £6 million for electrostatic precipitators, £3 million for a tall chimney and £6 million for ash-handling facilities. About £20,000–25,000 are dedicated to noise control. If needed, cooling towers average £6–10 million/plant. Air pollution monitoring equipment costs £50,000–60,000 per plant for 10–12 instruments. Detailed costs for 1990 are provided in Appendix C [Barrett 1982].

Currently, pollution control costs are not required to be reported in a systematic and vigorous manner. In its Fifth Report, the UK Royal Commission on Environmental Pollution felt that a weakness of the Inspectorate was that they "have neither accounting expertise to enable them to assess a company's financial situation nor economic expertise to consider the economic costs and benefits to the nation." (At the same time, the Royal Commission made it clear that this did not mean that the Inspectorate's decisions had been unsound in the past.) In practice, in making their economic assessment of a firm's position, the Inspectorate takes into account factors such as the industry's economic position, its significance in the economy, its recent financial record and the employment effects of any decisions on pollution control. However, costs and other economic considerations are not set out formally in comparison to technical data [OECD 1979a].

For the country as a whole, about 13 and 87% of the total annual investments on pollution control for all industries are associated with controlling air and water emissions, respectively. This ratio represents an estimated average over the period 1971–1980. The long-term (1960–1980) average of all pollution control investments as a percentage of the total projected investments for all industries in the country is about 0.9% [OECD 1974].

UNITED STATES

The U.S. electric utility industry spends a greater amount of money on antipollution equipment than any other industry. In 1980 the industry invested $2.6 billion in pollution control equipment. This amount represented slightly more than 28% of the combined expenditures of all businesses on similar investments [McGraw-Hill 1981a,b].

In addition to the capital costs of installing abatement facilities, which account for about 10% of annual capital expenditures, electric utilities have substantial operating costs and reliability problems associated with the installed equipment. For example, forced-air water cooling towers must use a portion of the electrical output to power fans that blow cooling air through the center section of the structure. This method of reducing thermal pollution can consume 3% of electrical output [Scott 1976].

Total Costs

Recently, the General Accounting Office surveyed four utility companies (Carolina and Florida Power and Light, Houston Lighting and Power, and the Cooperative/United Power Associations of Minnesota) to obtain information on costs and operational problems related to regulatory compliance. At the four utilities an estimated $1.4 billion in capital investments was needed to comply with regulations promulgated by eight federal and various state agencies. In addition, the utilities estimated that regulatory compliance would result in over $109 million in annual recurring costs (as estimated in 1980 dollars). More than 92% of the capital costs to date resulted from compliance with regulations promulgated by the U.S. Environmental Protection Agency (EPA) (73%), the Nuclear Regulatory Commission (NRC) (7%), and the states (12%). About 82% of the capital costs to date were associated with protection of the environment and wildlife, 17% with protection of the public interest and safe plant operation, 1% with monitoring financial and technical operations, and less than 0.1% on the protection of worker health, safety and job status [GAO 1981b].

Capital Investments

During the period 1970–1980, investor-owned electric utilities spent $18.1 billion (21.4%) on pollution control facilities, out of $84.3 billion spent by all U.S. business. The closest industry is petroleum, with an expenditure of $8.3 billion (9.8%) for the same period. During this time, investor-owned electric utilities spent a reported $11.4 billion in capital expenditures on air pollution control, $5.8 billion on water pollution control and $860 million on solid waste control (since 1975). In proportion, these expenditures for pollution control by the electric utilities are 63% for water pollution control, 32% for water pollution control and 5% for control of solid wastes [McGraw-Hill 1981b].

The data for electric utilities are presented in Table 3-15. The data indicate an almost threefold relative increase in the amount of pollution control expenditures as a percent of capital spending during the period. The methodology used for the survey has not been revealed by the company; thus, it is difficult to compare with other estimates. However, the survey and analysis of its results are sponsored by U.S. industries and used by the industries in evaluating trends in pollution abatement expenditures.

The Bureau of Economic Analysis, U.S. Department of Commerce, also publishes the results of an annual survey of business investments in pollution control. One of the drawbacks of the survey is that it samples company enterprises, not actual plants. A company is categorized by its major product. For example, if a large petroleum company owns a textile mill, any investment at the mill would be classified under petroleum refining. Thus, while the aggregate results of the survey are useful in assessing trends in pollution abatement costs, some individual industry-specific estimates reveal little about the cost to a specific industrial process. For the electric utility category, no differentiation is made among expenses associated with the operation of an electric utility or expenses with other activities of the utility, i.e., in some cases gas and water utility operation [EPA 1979].

In general, the amount of total pollution abatement costs and those for air, water and solid wastes follow the same trend as those estimated by McGraw-Hill, but vary from 9 to 13% higher than the McGraw-Hill estimates. One interesting analysis prepared by the Bureau of Economic Analysis is a calculation of planned new plant and equipment expenditures for pollution abatement as a percentage of actual expenditures. For electric utilities, the percentage decreases steadily from 108.5% in 1974 to 93.4% in 1980. The decrease is most striking for solid waste control expenditures, for which planned expenditures exceeded actual expenditures until 1980, when the percentage changed from 103 to 70.6%. The trend in the ratio of planned to actual expenditures may indicate a reluctance of the electric utility industry to invest in capital improvements in pollution control equipment until relevant pollution control regulations are promulgated, but funds are set aside as a contingency. Recent expenditures in excess of earlier planned expenditures may be anticipation of more stringent regulations that may affect future plants currently under construction [Rutledge and O'Connor 1981].

Two other surveys were conducted on pollution control investments. One was conducted by the Bureau of the Census, U.S. Department of Commerce, and the other by the Business Roundtable, a group of 48 of the nation's largest companies. Both of these surveys have limited sample sizes and deal mostly with manufacturing industries. The electric utility industry is excluded.

EPA has sponsored various macroeconomic model projections of incremental pollution control costs. Incremental costs exclude expendi-

Table 3-15. Capital Investments in Pollution Control Equipment for Electric Utilities in the United States[a] [McGraw-Hill 1981b]

Pollution Control Category	1970	1971	1972	1973	1974	1975	1976	1977	1978	1979	1980
Air	253	319	856	795	715	999	1462	1579	1188	1808	1394
Water	148	221	218	345	466	400	507	689	883	1006	947
Solid Waste						116	187	88	72	145	252
Total	401	540	1074	1140	1181	1515	2156	2356	2143	2959	2593
Total Pollution Control Expenditures as a Percent of Capital Spending	3.8	4.4	7.9	7.6	7.1	9.1	11.5	10.6	8.7	10.7	9.2

[a]Costs are in millions of dollars. Data for solid waste were included starting in 1975.

tures for pollution abatement, which, it is assumed, would be made in the absence of federal environmental legislation. The projections are forecasts of future anticipated expenditures required to comply with existing legislation and are not a retrospective compilation of actual costs. The results of the forecasts have been summarized in annual reports of the Council on Environmental Quality [CEQ 1978,1979,1980].

The macroeconomic forecasts are intended to predict major effects of legislation on the costs of future pollution control expenditures. For the electric utility industry the forecasts address such effects in general terms. The forecasts do not recognize costs involved in the loss of capacity during installation of pollution control equipment, and underestimate the high costs of low-sulfur coal and/or the actual cost of pollution control equipment. The forecasts do not consider the regulatory requirements of some states, which are more stringent than federal programs [Cohn 1975].

The most current of the macroeconomic forecasts indicates that utilities (electric, gas, water) will have spent $101 billion on the purchase and O&M of pollution control equipment because of federal air and water pollution control regulations between 1970 and 1987. This amount represents about 20% of similar expenditures by all businesses during the same period. Of the $101 billion, $55 billion will have been spent on O&M costs and $46 billion on the purchase of pollution control equipment. All costs are expressed in 1979 dollars [Data Resources, Inc. 1981; McGraw-Hill 1981].

The most reliable data available on the costs of pollution control associated with electric power plants in the United States are costs reported to the Federal Energy Regulatory Commission. The data are submitted within the commission's Uniform System of Accounts, which assures uniformity of treatment of company statements. The data are compiled from the individual utilities that comprise nearly 100% of the privately owned sector of the electric light and power industry. Data detailing investments in environmental protection facilities and actual expenses incurred in environmental protection have been reported, beginning with 1976. The data submitted to the Federal Energy Regulatory Commission are summarized and published by the Energy Information Administration of the U.S. Department of Energy [DOE 1977,1978, 1979,1980,1981a].

The reported data reflect recorded electrical plant in-service capital costs applicable to environmental protection facilities. The capital costs are summarized according to facilities dedicated to air pollution control, water pollution control, solid waste disposal, noise abatement, esthetic values, additional plant capacity and miscellaneous protection facilities.

Other reported data reflect actual operating expenses related to

Table 3-16. Capital Costs of In-Service Pollution Control Facilities in the United States[a] [DOE 1977,1978,1979,1980,1981a]

Focus of Environment Expenditures	1976	1977	1978	1979	1980	Total Over 5-Year Period	Percent of Total Environmental Expenditures over the 5-Year Period
Air Pollution	3,436	4,527	6,147	7,147	8,993	30,250	43
Water Pollution	1,371	1,756	2,410	2,898	3,753	12,188	17
Solid Waste Management	528	616	839	922	11,228	4,133	6
Noise Control	136	155	166	183	197	837	1
Protection of Esthetic Values	2,685	3,100	3,609	4,109	4,593	18,096	26
Miscellaneous	666	800	1,041	1,116	1,740	5,363	7
Total	8,822	10,954	14,212	16,375	20,504	70,867	100
Percentage of Total Capital In-Service for Given Year Attributable to Environmental Expenditures	6.2	7.0	8.4	9.0	10.3		

[a] Costs are in millions of dollars in currency of the year given. Data represent worth of environmental facilities as reported by U.S. privately owned electric utilities. Data begin with the first year data were compiled.

environmental protection facilities. These data are not summarized according to control categories, as are the reported capital costs. They are broken down by environmental protection costs incurred as depreciation; labor, maintenance,material and supplies; fuel-related costs; replacement power costs, taxes and fees, administration, and miscellaneous activities.

The capital costs of in-service pollution control facilities in the United States are presented in Table 3-16. The amount of capital costs of operating environmental protection facilities at U.S. electric power stations has increased substantially since the data were first reported in 1976. The percent of total capital in-service for a given year that was attributable to environmental protection has risen from 6.2% in 1976 to 10.3% in 1980. During the period 1976–1980, 43% of environmental protection capital in-service costs was dedicated to air pollution control. The second highest percentage (26%) of in-service capital costs associated with environmental protection was attributable to the protection and enhancement of esthetic amenities. Water pollution control facilities accounted for 17% of the in-service environmental protection costs. Solid waste management and miscellaneous protection facilities accounted for 6 and 7%, respectively, of the total in-service capital costs associated with environmental protection.

Operating and Maintenance Costs

Operating and maintenance costs related to environmental protection facilities at privately owned U.S. electric utilities are presented in Table 3-17. The total operating and maintenance costs associated with the plants have averaged 2% over the period 1976–1980.

Cost Recovery

In a response by public utility commissions in 49 states and the District of Columbia to a 1980 survey, 33 said they allow either all or part of construction work in progress in the rate base. Of these 33, 28 allow pollution and environmental control projects to be included as part of the rate base attributed to construction work in progress. This represents preferential treatment for recovery of environmental expenditures, as other types of expenditures were less favored by the commissions. However, while some state utility commissions allow utilities to include plant construction costs in their rate bases as these costs are incurred, other commissions only allow some of the construction costs, and still other commissions do not permit utilities to begin recovering construction costs until new plants are put into operation (Table 3-18 [GAO 1980b,c].

Table 3-17. Operating and Maintenance Costs Related to Environmental Protection Facilities at U.S. Investor-Owned Electric Utilities (Thousands of Current Dollars) [DOE 1977,1978,1979,1980,1981a]

Year	Costs	Percent of Total O&M Costs
1970–1975	Data Not Reported	
1976	737,811	1.8
1977	985,482	2.0
1978	884,080	1.6
1979	2,227,730	2.1
1980	1,618,797	2.2

Table 3-18. Types of Construction Work in Progress Allowed in the Rate Base by State Commissions[a] [GAO 1980b]

Type of Construction-Work-In-Process (CWIP)	Number of Commissions
Production	26
Transmission	25
Distribution	24
General Plant	24
Pollution/Environmental Control Project	28
Fuel Conversion Projects	16
Nuclear Plant	18
Other	6

[a] Some commissions have not allowed some types of CWIP, such as fuel conversion projects or nuclear plant, in the rate base, either because none exists in their states, or these types have not been addressed in a rate case.

Effect of Pollution Control Regulations on the Reliability of Electric Service

Compliance with environmental regulations can conflict with the basic responsibility of electric utilities to provide reliable power at reasonable rates. Reliability of power can be affected by the inability to build new generating capacity, lengthy delays in completing construction projects and diversion of power from customer service to operate pollution control equipment.

An important aspect of reliability relates to the "capacity penalty," which is the percent of generation used to power controls. For a new plant

Table 3-19. Average U.S. Power Plant Delays, 1974-1978 [GAO 1980c]

	1974	1975	1976	1977	1978	Total
Nuclear						
Units Delayed	11	9	4	7	3	34
Average Delay (months)	26	28	27	45	59	33
Coal						
Units Delayed	8	6	10	16	15	55
Average Delay (months)	5	5	16	9	11	10
Oil						
Units Delayed	7	9	5	9	5	35
Average Delay (months)	12	10	12	19	24	15
Other						
Units Delayed	7	5	4	4	5	25
Average Delay (months)	9	10	14	15	35	16
Total						
Units Delayed	33	29	23	36	28	149
Average Delay (months)	14	14	16	19	23	17

with stringent controls, this penalty is about 5%. This results in less efficient generation overall and in substantial costs to the consumer. For example, in a study of the Carolina Power and Light Company, the O&M cost for electrostatic precipitators in 1977 was more than $1.4 million, of which about $900,000 (64%) was the cost of power to operate the equipment [GAO 1981b].

Reliability also can be affected by delays in power plant construction. A delay is a slippage, or projected slippage, beyond the initial projected date of operation. Utility companies have reported to the General Accounting Office that most of the costs of protecting the public are attributable to construction delays caused by lengthy and complicated federal and state regulatory processes, although changes in consumption patterns within the last few years are partially responsible for some power plant cancellation and delays [GAO 1981a,b].

Since 1974, most new generating units have been delayed. During the period January 1974 through December 1978, 189 electric generating units were put in operation. Of these, 149 units (79%) experienced delays of less than 6 months to more than 3 years, averaging 17 months. As of December 1980, another 330 units were planned to be completed by the early 1990s. Of these, 267 (81%) were already delayed an average of 50 months [GAO 1980c].

The length of delays is increasing each year. For example, an average delay of 14 months was experienced by 33 generating units that came on line in 1974 (Table 3-19). The 28 units starting operation in 1978 were delayed by an average of 23 months. Nuclear plants have incurred the longest delays, averaging 33 months for 1974–1978, compared to an average of 10 months for coal units. The total times required for power plant completion in 1974–1975 including delays were 78–90 months for fossil-fueled plants and 120 months for nuclear power plants (Table 3-20). These times are much longer than those required for other complex energy projects, such as refineries, which averaged 60–78 months [GAO 1980c; OECD 1977].

In addition to delays, power plant cancellations also can affect reliability of service. During the period 1974–1978, the U.S. electric utility industry cancelled 184 planned, large electric operating units including 80 nuclear and 84 coal-fired plants. The capacity of these plants would have been equivalent to about 26% of the existing total generating capacity as of April 1979 [GAO 1980c].

Effect of Environmental Protection Laws on Power Station Costs

Electric utilities earn a fair and reasonable rate of return on costs incurred in the construction of power plants. These costs are borne by the consumer. When new legislation, such as amendments to the Clean Air Act, set new requirements for stack emissions and power plants, utilities may be forced to make additional investments on existing power plants by retrofitting pollution control equipment on those plants. Since these old plants have been used continually and for the most part depreciated, such new regulations can have the effect of requiring companies to increase their capital investment in the old plants by as much as 45% and operating cost by as much as 100%. With respect to new facilities, in 1979 about 25% of new coal-fired plant construction costs were for environmental equipment designed to bring the new facilities into compliance with regulatory standards [Elkin 1979].

The two major reasons for increased power station costs are inflation and regulatory compliance. Both factors escalate costs with time and, when plant completion is delayed pending permit acquisition, the final plant cost can be increased substantially as a result. Of the 10 years needed to build a coal-fired power plant in 1979, only 3 years were required for the actual physical construction while 7 years were required to obtain permits necessary to build the plant. For a nuclear plant, 14 years were required for completion, if all of the needed permits could be acquired [Elkin 1979].

The cost to construct coal-fired electric generating plants in the United States increased significantly during the 1970s. Although inflation in construction wages and material prices was a contributory factor, increased environmental standards played a particularly important role. A

Table 3-20. Times Required for Completion of U.S. Electric Power Plants[a] [OECD 1977]

Activity	Fossil-Fueled Power Plant (months)	Nuclear Power Plant (months)	Oil Refinery (months)
Site Selection	12	18	12
Site Analysis	24	24	12–18
Public Hearings[b]	12	12	6
Regulatory Agency Review and Decision[b]	6	6	6
Construction	24–36	60	24–36
Total	78–90	120	60–80

[a] All of the time periods indicated in the table were typical for expeditious permitting of facilities in construction in 1974–1975. Each period can vary significantly for particular facilities.

[b] These represent minimum times and often result in prolonged delays for environmental review.

statistical analysis of the capital costs of recently completed U.S. coal-fired generating units indicated that the cost to build a typical coal plant increased by 69% from the end of 1971 to the end of 1978, in addition to inflation in construction labor and materials. Approximately 90% of this real-dollar increase was spent for improvements in pollution control. In return, emissions of sulfur oxides, particulates and nitrogen oxides from 1978 plants averaged approximately 64% less than those from 1971 plants [Komanoff 1980].

The use of coal to generate electricity expanded rapidly in the 1960s and 1970s. U.S. coal-fired generating capacity increased by 80% from 1961 to 1971 and by another 53% to 1978. The resulting increase in coal-generated emissions provoked national concern, inspired by massive new pollution sources such as the Four Corners plant in northern New Mexico. Accordingly, starting in the mid-to-late 1960s, some state and local authorities ordered utilities to reduce emissions of particulate matter and sulfur dioxides through fuel switching or improved control devices. In 1970 Congress passed amendments to the Clean Air Act that created a framework for reducing emissions from existing plants and set national standards for new plants [Komanoff 1980].

The new source performance standards (NSPS) promulgated by the EPA pursuant to the amendments limited emission of particulates, sulfur dioxide and nitrogen oxides from fossil-fueled plants whose construction started after August 1971. The standards required emissions of these pollutants from new coal plants to be 55% less per unit of fuel burned, on average, than emissions from plants installed in 1971. Some new plants have surpassed the levels, as a result of stricter local regulations, state measures needed to satisfy national ambient air quality standards or utility efforts to keep ahead of regulations [Komanoff 1980].

The result of the regulations has been an increase in the average cost to construct coal plants from $346/kW of capacity for late 1971 completions to about $583/kW for late 1978 plants, as measured in constant mid-1979 dollars. About 90% of the increase ($210–215/kW) is attributable to the cost of required pollution control activities. A breakdown of environmental control factors that have contributed to increased power plant costs is provided in Table 3-21.

Table 3-21. Increases in Pollution Control Costs for New Coal-Fired Power Stations, 1971–1978 (1979 $/kW)[a] [Komanoff 1980]

Pollutant/Activity	1971	1978	Reasons for Increase
Sulfur Dioxide		120	Average cost of scrubber including sludge handling and disposal (fixation and ponding) systems to comply with new source performance standards of the Clean Air Act
Particulate Matter	20	60	Average cost for increase in efficiency in electrostatic precipitators from 97 to 99.5%
Nitrogen Oxides		10	Boiler modifications resulting in a 35% lower emission rate
Noise		10	Noise attenuation features
Construction Pollutants		5	Abatement during plant construction
Liquid Waste	2	10	Systems to treat normal plant waste drains for reuse or for discharge
Solid Waste	0–5	5	Improved ash disposal techniques
Air Pollutants	1	2	Air pollution monitoring systems
Heat	2	5	Average cost of increased use of cooling towers
Boiler Improvements		10	Needed to accommodate variations in coal grade caused by mine safety rules
Data Reporting		3	Preparation of environmental reports to state and federal agencies
Total	25–30	240	Resulting in the average cost to construct coal plants from $346/kW of capacity for late 1971 completions to $586/kW for late 1978 plants, a 69% increase in addition to inflation in construction labor and materials.

[a] Data based on costs of all 116 U.S. coal plants over 100 MW completed during 1972–1977. It was assumed that a 1978 plant includes a scrubber to remove sulfur dioxide, although about half of recent coal plants lack scrubbers because of using low-sulfur coal to comply with new source performance standards. All costs are in constant mid-1979 dollars and have been adjusted to reflect prevailing prices of labor, materials and equipment in 1979.

Effect of Pollution Control Requirements on the Costs of Future Power Stations

U.S. environmental legislation in the 1970s will markedly increase the cost of power plants completed in the 1980s. Regulations promulgated under the Resource Conservation and Recovery Act (RCRA) and amendments to the Clean Water and Clean Air Acts require new or more refined pollution control equipment as well as additional monitoring and data reporting requirements (Table 3-22).

During the 1970s, the percent of total in-service capital investment dedicated to pollution control activities reached a high of 9%. The proportion of capital cost for a typical new coal-fired power station completed in 1980 that was attributable to environmental control requirements was about 30–35%. By 1985 potentially more stringent requirements are expected to increase environmental control costs to 50–55% of total capital costs [EPRI 1980a].

In addition to capital investment costs, pollution control requirements will also result in higher costs for maintenance of pollution control equipment and for monitoring. Under the NSPS program, EPA requires electric utility steam generating units to install and operate continuous in-stack monitors (Table 3-23). In addition, 37 states and local agencies have included some specific provisions for continuous, in-stack monitoring in their state implementation plans (Table 3-24).

Specific Effects of Current U.S. Regulations on the Cost of Future Power Plants

Most of the information published with respect to pollution control costs is devoted to forecasting particular effects of existing or proposed regulations on the costs of the construction and operation of future facilities. Although a review of this information is not a focus of this chapter, major increases in the costs of recent power plant construction work in progress are associated with the need to meet more stringent regulations that are being implemented or will be implemented in the near future. The following are a few examples of such forecasts.

Water Intakes

Section 316(b) of the Federal Water Pollution Control Act addresses the location, design, construction and capacity of cooling water intake structures, and requires that such structures use the best available technology to minimize adverse environmental impact. Although the

Table 3-22. Evolution of U.S. Coal-Fired Power Plant Environmental Control Requirements [EPRI 1980a]

Pollutant or Construction Requirement	Type of Environmental Requirement		
	1969	1978	1985
Particulate Matter	95% removal	More than 99% removal	At least 99.9% removal
Sulfur Dioxide	Percent sulfur limit on fuel	Nonregenerable flue gas desulfurization	Implementation of regenerable flue gas desulfurization
Nitrogen Oxides	Control required only in California	Combustion control	Implementation of post-combustion control
Water Pollutants	Oil separation	Single discharge	Use of wastewater treatment to attain "zero" discharge
Solid Waste	Ponding	Inventory and offsite disposal	Long-term monitoring by product recovery
Noise	No special requirement	Attenuation	Equipment standards
Thermal Discharge	Open cycle, usually once-through cooling	Cooling tower	Wet or dry cooling towers as appropriate
Auxiliary Load	1%	10%	30%
Licensing	Limited environmental review	Environmental impact statement and hearings	Preconstruction review
Construction Schedule	4 years	7 years	?

Table 3-23. New Source Performance Standards Requirements for the Continuous Monitoring of Air Emissions from Electric Utility Steam-Generating Units [OTA 1982]

	Purpose of Monitoring
Solid or Liquid Fossil Fuel	
Opacity	Verification of proper operation and maintenance of process and control equipment
Sulfur Dioxide (at inlet and outlet control device)	Performance (compliance) test
Nitrogen Oxides Oxygen or Carbon Dioxide	Performance (compliance) test
Gaseous Fossil Fuel	
Nitrogen Oxides	Performance (compliance) test
Oxygen or Carbon Dioxide	Performance (compliance) test

Table 3-24. State Implementation Plans: Requirements for the Continuous Monitoring of Air Emissions from Fossil Fuel-Fired Steam Generators [OTA 1982]

Pollutant	Conditions That Require Monitoring
Sulfur Dioxide	Plant processes more than 250 million Btu/hr or source employs sulfur dioxide control equipment
Nitrogen Oxides	Plant processes more than 1 billion Btu/hr or located in a designated nonattainment area for nitrogen oxides; source exempt if is at least 30% below emission standard
Opacity	Plant processes more than 250 million Btu/hr; source exempt if burns gas, oil or oil mixture that results in compliance for particulate matter and opacity without control equipment

section relates to all industries, about 80% of all U.S. industrial cooling water intake is used by steam-electric power plants.

Regulations implementing the section require case-by-case determinations in which environmental effectiveness must be examined in light of economic considerations. Options that need to be examined include conventional vertical traveling screens, micromesh screens, and reduction of intake cooling water volume and velocity.

Although the costs and economic effects of control options will vary with local plant conditions, it is determined that, with the exception of reducing cooling water intake volume, the control options will add little to the costs of producing electricity at a power station (i.e., increasing the base production cost by 0.01–0.76 mill/kWh). Use of the reduction of cooling water volume option will likely increase base costs by 3.3 mill/kWh. Overall, compliance with section 316(b) regulations will likely require total capital expenditures of about $895 million by the electric power industry through 1987, and increase annual operating expenses by about $100 million in 1987, resulting in a 0.2% increase in expected operating revenues (all costs are in 1977 dollars) [Temple, Barker and Sloane, Inc. 1977].

Air Emissions

A revised set of NSPS, approximately twice as stringent as the original standards, was promulgated in 1979 pursuant to the Clean Air Act Amendments of 1977. These new standards pertain to plant construction begun after September 18, 1978, and may affect plants that come on line as early as 1982.

The 1977 amendments also require that new utility and industrial plants built in or near designated pristine [prevention of significant deterioration (PSD)] areas or polluted (nonattainment) areas install the "best available control technology" or achieve the "lowest achievable emission rate," respectively. These guidelines are defined, ambiguously, as the maximum reduction possible for each pollutant, taking into account energy, environmental and economic impacts. They are intended to be "technology-forcing," i.e., to push the utility industry to develop improved controls surpassing the new standards. The actual reductions required are to be determined on a site-specific basis in "new source reviews" performed during the permitting process [Komanoff 1980].

The new standards will thus serve as a floor, rather than a ceiling, to pollution control practice for many new coal plants. Over half of the country either is in a PSD or nonattainment area or will affect such an area via plume transport. Thus, a majority of new plants may be required to improve the NSPS. Some utilities may opt for stricter controls to avert drawn-out negotiations. The NSPS are subject to further strengthening as coal-fired generating capacity continues to expand. The prohibition of new oil- or gas-fired generators and the diminished prospects for nuclear power assure that coal plants will provide more than half of the capacity increase required through most of the 1990s [Komanoff 1980].

Conservative cost estimates for pollution control improvements to meet the new standards are estimated to be about $190/kW of capacity (in 1979 dollars) during the period 1978–1988. This amount is almost equal to the cost of the controls added during the period 1971–1978. A breakdown of environmental control factors that are expected to contribute to the increase in power plant costs in the future is provided in Table 3-25.

Table 3-25. Increases in Pollution Control Costs for New Coal-Fired Power Stations, 1978–1988 (1979 $/kW)[a] [Komanoff 1980]

Pollutant/Activity	1978	1988	Reasons for Increase
Sulfur Dioxide	120	140–180	A 90% reduction in sulfur dioxide emission is needed to meet a 30-day continuous averaging requirement; costs may be higher to attain 95% removal efficiency for acid rain control
Particulate Matter	60	65–80	Increase in collection efficiency from 99.1 to 99.9%
Nitrogen Oxides	10	60–90	Need for monitoring and control systems, corrosion-resistant materials and catalytic reduction
Solid Waste	5	30–45	Lining of sludge and ash holding ponds and use of regenerable systems which recycle waste products
Other	45	65–75	Near-zero water discharge, noise attenuation, dry cooling, stringent air and water monitoring and data reporting requirements
Total	240	360–470	The total increase in the capital costs of environmental controls for a typical 1988 plant, compared to 1978, could range $120–230/kW in addition to the inflationary costs of construction labor and materials

[a] Data are projected from reported costs of all 116 U.S. coal plants over 100 MW that were completed during 1972–1977. The increased costs are largely due to compliance with regulations implementing the Clean Air Act Amendments of 1977 and the Resource Conservation and Recovery Act of 1976. All costs are in constant mid-1979 dollars and have been adjusted to reflect prevailing prices of labor, materials and equipment in 1979.

CHAPTER 4

RESEARCH AND DEVELOPMENT COSTS

TOTAL EXPENDITURES

The Federal Republic of Germany planned to devote about 0.1% of its gross national product (GNP) to research and development (R&D) of pollution control equipment during the period 1971–1975. This amount was expected to be spent by industry in compliance with a developing national environmental policy [OECD 1972].

On a national scale, a commonly used indicator is R&D expenditure as a percentage of GNP. For the most industrialized nations, this ratio ranges from 2.0 to 2.5%. In the 1970s, however, the Federal Republic of Germany's R&D investments were increasing while U.S. investment was decreasing. U.S. federal R&D obligations, while increasing in current dollars, declined in constant dollars to a low in 1975. Only in the late 1970s did they regain their earlier level. While federal funds decreased, industrial support of research increased [EPRI 1980b].

Great Britain's Central Electricity Generating Board (CEGB) and Electricité de France (EDF) have both increased the ratio of R&D expenditures to sales, the 1979 levels being 1.3 and 2.0%, respectively (1981–1982 CEGB environmental research is expected to be about 9% of total research costs). Although it is difficult to compare foreign nationalized utilities, such as CEGB and EDF, with U.S. privately owned utilities or even all U.S. utilities, the differences may be indicative. Coupled with the fact that U.S. electric power utilities collectively spend less on R&D than do most technologically based U.S. companies, one is left with the impression that the current electric utility R&D level is low compared to industries. On a national basis, industry has been investing in research at a rate of about 1.9% of sales, a rate that has been constant despite variations in profit levels and economic conditions. In the manu-

Table 4-1. Costs of Research and Development Related to Pollution Control Activities at Electric Power Stations in France (1970–1980) Broken Down by Pollution Control Category and by Nature of Study (Millions of 1980 tax-free francs) [EDF 1982]

Type of Research and Development	1970	1971	1972	1973	1974	1975	1976	1977	1978	1979	1980
Pollution Control Category											
Air	6.7	10.4	13.6	15.5	20.5	17.6	32.7	31.6	41.1	35.4	32.3
Water	3.2	2.2	11.0	14.6	27.4	41.8	72.7	65.9	62.2	65.5	73.3
Esthetics						0.3	6.6	7.1	6.0	2.1	16.2
Radiation					0.3	1.1	7.5	8.2	8.3	4.1	16.4
Noise	2.7	3.8	6.1	4.2	9.6	9.9	13.2	16.7	19.6	19.6	20.0
Total	12.6	16.4	30.7	34.3	57.8	70.7	132.7	129.5	137.2	126.7	158.2
Nature of Study											
Basic Research	12.6	16.4	30.7	34.3	53.8	59.7	59.9	69.1	76.9	76.6	85.6
Equipment Development					4.0	11.0	63.7	51.4	53.1	42.0	65.3
Process Improvement							9.1	9.0	7.2	8.1	7.3
Total	12.6	16.4	30.7	34.3	57.8	70.7	132.7	129.5	137.2	126.7	158.2

facturing industry, research support averages slightly over 2%. The range is from 0.2% (food and beverages, textile) to more than 20% (aerospace). The electric utility industry stands at about 0.65% of sales [EPRI 1980b].

FRANCE

As cited earlier, environmental legislation allows investment in air or water pollution facilities and equipment to qualify for accelerated depreciation. In addition, the Ministry for Industrial and Scientific Development may aid industry by providing 50% of the costs to develop new products or processes destined for pollution abatement. However, the grant must be reimbursed if the product or process becomes commercially feasible [Congressional Research Service 1981].

About 50% of pollution control R&D sponsored by electric power stations in France is dedicated to water pollution control, about 20% to the abatement of air pollution, and the rest devoted to noise control, radiation and esthetics. About 54% is basic research. The development of new or refined equipment accounts for 41% of all research expenditures. About 5% deals with process improvements (Table 4-1).

UNITED STATES

Total R&D Expenditures

Based on 1980 expenditures by investor-owned utilities, about 50% of R&D and demonstration activities are performed or funded directly by the utilities. The trend over the last several years has been to increase this percentage by about 1% each year; that is, relatively more R&D is being managed each year within the companies [DOE 1981b].

Of the 50% of R&D and demonstration funds borne by the utilities, but expended by other institutions, half are used to provide research support to the Electric Power Research Institute (EPRI). The remainder is used to support the Edison Electric Institute (EEI) which receives about 26% of the external funds, various nuclear power research groups adjunct to EPRI, and various other small research entities. Funds are derived from utilities in proportion to their electricity sales and are coordinated and distributed through two trade associations: the American Public Power Association and the National Rural Electric Cooperative Association. Table 4-2 presents a breakdown of the total research development

Table 4-2. Expenditures by Investor-Owned Electric Utilities for R&D and Demonstration [DOE 1981a]

	1979		1980	
	Amount ($ millions)	Percent of Total	Amount ($ millions)	Percent of Total
Power Plants				
Hydroelectric				
Recreation, fish and wildlife	1.3	0.3	0.7	0.1
Other hydroelectric	2.3	0.5	1.5	0.3
Fossil fuel steam	39.9	9.3	72.3	14.5
Internal combustion or gas	1.6	0.4	0.4	0.1
Nuclear	24.0	5.6	18.9	3.8
Unconventional generation	36.3	8.4	43.1	8.6
Siting and heat rejection	9.7	2.3	6.5	1.3
System Planning, Engineering and Operation	9.2	2.1	16.5	3.3
Transmission				
Overhead	5.4	1.3	4.0	0.8
Underground	1.3	0.3	1.2	0.2
Distribution	5.7	1.3	8.5	1.7
Environment	40.6	9.4	44.3	8.9
Other	32.2	7.5	30.5	6.1
Total Within Utilities	209.5	48.7	248.4	49.7
Research Support to				
Electric Power Research Institute	111.1	25.8	122.4	24.5
Edison Electric Institute	63.4	14.7	67.1	13.5
Nuclear Power Groups	12.0	2.8	9.5	1.9
Others	34.6	8.0	52.0	10.4
Total Outside Utilities	221.1	51.3	251.0	50.3
Total	430.6	100.0	499.4	100.0

expenditures by privately owned electric utilities for the years 1979 and 1980 [DOE 1981b; Lawrence 1982].

EEI is an association of investor-owned electric utility companies. It gathers information and statistics relating to the electric power industry and makes them available to member companies, the public, and state and federal agencies. EEI also maintains liaison between the industry and the federal government and acts as a voice for the electric utility industry on subjects of national interest.

EPRI, formerly the Electric Research Council affiliated with EEI, develops and manages a technology and environmental research program dedicated to improving electric power production, distribution and utilization. EPRI also manages four separately funded research efforts. These are the Steam Generator Owners Group Program, the Boiling

Table 4-3. In-House Environmental R&D and Demonstration Expenditures of U.S. Investor-Owned Electric Utilities [DOE 1977,1978,1979,1980,1981a]

Year	Expenditures ($ millions)	Percent of Total
1973	6.9	5.0
1974	18.6	17.0
1975	27.3	24.2
1976	24.5	19.7
1977	30.2	18.1
1978	39.9	20.8
1979	40.6	19.3
1980	44.3	17.9

Water Reactor Owners Group Intergranular Stress Corrosion Cracking Program, the Pressurized Water Reactor Safety and Relief Valve Program and the Nuclear Safety Analysis Center [EPRI 1981b].

Environmental R&D Expenditures

While capital and operating costs account for the major share of funds dedicated to environmental control, electric utilities also spend money on research in the area of pollution abatement. In 1980, investor-owned utilities spent about $500 million on pollution control R&D or about 25% of the total pollution control R&D expenditures by U.S. industries [DOE 1981b; McGraw-Hill 1981a,b].

Some R&D funds are spent by utilities on in-house environmental research activities. Over the last 8 years this portion has averaged about 18% of all funds devoted by the electric utilities to environmental R&D (Table 4-3).

Since the utilities do not have extensive research facilities, but share common R&D needs, economy of scale and efficiency are attained by shared funding of research. Research funds collected from participating electric utilities are pooled to support a comprehensive research and development program, largely directed and implemented by EPRI. Within the area of environmental assessment, EPRI currently emphasizes research required to improve understanding of the effects of air and water effluents, electric fields, and solid wastes. EPRI also maintains an equally strong program focused on the development of the control technology needed to maintain potential environmental effects at acceptable levels.

Environmental assessment research is comprised of programs in environmental physics and chemistry, environmental risk and issues analysis, ecological studies, and environmental and occupational health. Pollution control technology has programs related to coal quality (with emphasis on coal cleaning), fluidized-bed combustion and alternative fuels (designed to reduce the cost of controlling emissions), air quality control, desulfurization processes, and water quality control and heat rejection [EPRI 1981b].

In 1981 EPRI devoted about 25% of its budget to environmental protection research associated with the production of electric power. This amount is expected to decrease to 23% during 1982 and to maintain this level throughout 1981–1985 (Table 4-4) [EPRI 1981b].

Table 4-4. Environmental Protection Research Expenditures Associated with the Production of Electric Power as Directed by EPRI (1981–1982 Planned Expenditures in $ Millions)[a]

Program	1981	1982
Combustion (Emphasis on Coal)		
Coal Quality	4.0	4.7
Fluidized-Bed Combustion and Alternative Fuels	6.9	7.5
Air Quality Control	9.0	9.5
Desulfurization Processes	8.2	1.01
Water Quality Control and Heat Rejection	6.2	7.2
Environmental Assessment		
Environmental Physics and Chemistry	11.3	11.9
Environmental Risk and Issues Analysis	1.6	1.7
Ecological Studies	4.8	5.9
Environmental and Occupational Health	7.9	8.4
Total Environmental Protection Expenditures	59.9	66.9
Total Institute Expenditures	245.0	291.0
Environmental Protection Expenditures as a Percent of Total Expenditures	25	23

[a] Data extracted from a listing of planned expenditures in a special report by EPRI [1981b].

APPENDIX A

ESTHETICS

The countries addressed in this book deal with the problem of power plant siting esthetic issues in much the same manner, although few laws specifically deal with the problem. Tables A-1 and A-2 provide detailed information concerning the degree to which esthetics is addressed during power plant siting and construction in the various countries. The information in this section is derived from a report of a committee on electric power of the Economic Commission for Europe [ECE 1977a]. The following is a summary of the salient points of the report as presented in Tables A-1 and A-2.

The types of power stations and the scale of size of the largest plants are essentially similar; thus, comparisons can be made with respect to the general problem of amenity. The common concerns focus on the height of the chimney and cooling towers as they affect the visual setting in which they are placed. These problems are difficult to avoid. In countries with extensive coastline and relatively short distances to inland population centers, estuaries or seawater can be used to dissipate waste heat in place of cooling towers (e.g., Northern Ireland). Where applicable, hydropower offers an alternative to the construction of tall chimneys and cooling towers in rural settings that have high esthetic value (e.g., Federal Republic of Germany).

All of the countries addressed have some form of system that enables consultation to take place, often involving the public, while a power station project is in the planning stage. Introduction of power stations with tall chimneys and cooling towers, large bulky buildings, and a concentration of transmission lines has created a visual impact that is highly noticeable to members of the public, especially those interested in preserving the amenity of a particular part of the countryside. Complaints

Table A-1. Comparison of Approaches to Minimize Harm to Esthetic Amenities Resulting from Power Plant Construction in the Federal Republic of Germany, France, the United Kingdom and the United States [Commission of the European Communities, 1979; ECE 1977a; OECD 1977]

Country	Nature of Public Involvement	Nature of Impact Mitigation	Areas Where Esthetic Amenities Preclude Construction	Additional Expenditures to Prevent or Reduce Harm to Esthetic Amenities
Federal Republic of Germany	Public relation activities are directed to ascertain public reaction to power plant proposals; siting issues are discussed with interested local and national authorities	Procedural rules are being prepared to deal with siting decisions; emphasis through 1977 has been on best outline design	A distinction is made between areas relating to the conservation of nature vs conservation of landscape and unprotected areas; Hydroelectric plants often are suitable for areas of high esthetic amenity	Additional costs are incurred from time providing the resultant scheme remains technically viable
France	Notices of power plant proposals are placed in the press and at town halls; local authorities and landowners affected are informed together with appropriate local and national government departments; recognized associations can plead under law before planning authorities; public inquiry largely limited to written comment and answer.	Although no specific esthetic rules exist on a national scale (as of 1977), use is made of siting, scale shape and color to harmonize an installation with the natural landscape; tree planting is encouraged.	No system exists (as of 1977) for classifying of land according to its esthetic value	Except for one case where public funds were used for cleaning up esthetically displeasing transmission networks, no special funds are set aside for the protection of amenities; emphasis is on environmental compatibility in siting and design of new power stations
United Kingdom				
Northern Ireland	Consultation with planning authorities and the Ulster Countryside Committee; when easements are sought,	No relevant rules exist, but early consultation with planning authorities is encouraged; emphasis is on use of natural	Areas of outstanding natural beauty, National Trust properties, forest parks, are examples of special esthetic	Unless compelled by planning authorities or the costs are recoverable, only marginal cost increases are considered for

	opportunity is provided for objections to be made	features to conceal the appearance of a plant and on tree planting; buildings usually required to be finished similar to others in the vicinity	amenity areas, but an absolute ban in these areas is not justifiable as electric power may have a higher degree of national importance	such purposes
England and Wales	Consultations with local planning authorities, Royal Fine Arts Commission and Nature Conservations; most involvement occurs at public hearings	Emphasis on use of natural features trees, and artificial mounds for screening, use of local stone for finishing of buildings, and on small clusters of cooling towers	Much of the country is designated as areas of outstanding natural beauty, green belt, great landscape value, historical or scientific, and national parks; no prohibition to construction exists, but National Parks may be prohibitions in reality	Increases in expenditure are incurred frequently
North and South Scotland	Consultation with local planning authorities, Secretary of State's Amenity Committee and the Royal Fine Arts Commission; local Amenity Associations are active participants	Emphasis is on use of walls and natural plantings for screening purposes	Same as above; a balance is struck on a site-by-site basis between the cost to the consumer as a user of electricity and also as a custodian of the environment	Additional costs are often incurred after many consultations to assess justification; essentially, planning boards are bound by law to try and preserve the beauty of scenery
United States	For nuclear and some nonnuclear facilities public involvement occurs in several stages in the development of an environmental impact statement; local hearings held by public utility regulatory commissions also offer an opportunity for public comment; citizen suits play an important part in environmental protection	Capital inservice cost for esthetics is the second highest (next to air pollution) capital cost for U.S. electric power stations; emphasis is on harmony with environmental and cultural surroundings	National Park and designated wilderness areas are generally considered to be areas that preclude power plant construction; the selection of a site within other areas is influenced a great deal by public and political pressure	Additional costs are made in compliance with relevant "esthetics" wording in several statutes, e.g., National Environmental Policy Act; funds are provided under law for citizens suits to force regulatory compliance

Table A-2. Comparison of the Treatment of Esthetic Issues Affecting Power Plant Construction in the Federal Republic of Germany, France, the United Kingdom and the United States [DOE 1981a; ECE 1977a; Ramsay 1979]

Country	General Types of Power Stations and Station Capacity	Notes on Esthetic Issues	Nature of Complaints Relating to Esthetics	Sources of Complaints
Federal Republic of Germany	Hydro up to 2110 MW, conv. thermal up to 2600 MW, nuclear up to 670 MW	Regulations govern the heights of chimneys	Complaints mainly relate to cooling towers or chimneys which are considered to be "overdimensionally too big"	Local inhabitants near the proposed power station and environmental preservationists
France	Conv. thermal up to 1800 MW, hydro up to 50 MW, nuclear up to 2000 MW	Tall chimneys and cooling towers	Power stations are rarely the subject of complaints; when they arise, most relate to cooling towers	Local bodies
United Kingdom				
Northern Ireland	Coal up to 240 MW, oil up to 1200 MW	Cooling towers not used since stations are situated on estuaries	Complaints generally relate to the overall impact; specific complaints cover the size and number of chimneys, oil storage, and the effect of noise	Local government bodies, countryside preservation committees, ecological interests, landowners and recreational interests
England and Wales	Coal up to 2000 MW, oil up to 2000 MW, nuclear up to 1320 MW, hydro up to 360 MW	Cooling towers are needed on most inland sites; very tall chimneys needed (about 260 m) for coal- and oil-fired plants	Objections are made to power station proposals, but objections are rare after construction is complete; those living close to proposed power stations generally comment on specific items, i.e., size of cooling towers	Local government bodies, countryside preservation committees, ecological interests, landowners and recreational interests

North and South Scotland	Coal up to 2400 MW, oil up to 2000 MW, nuclear up to 300 MW, hydro up to 123 MW	Most major plants are on the coast and cooling towers not needed; tall chimneys needed to disperse gases	Objections are made to power station proposals, but objections are rare after construction is complete; those living close to proposed power stations generally comment on specific items, i.e., size of cooling towers	For south Scotland, same as above; for north Scotland, mainly amenity associations, land developers, land-owners, and tenants
United States	Conv. thermal up to 2268 MW, hydro up to 6150 MW, nuclear up to 2300 MW, geothermal up to 985 MW	Cooling towers and tall chimneys	Complaints focus on visual impacts and influx of construction workers	Most complaints are from local inhabitants and ecological interests

concerning these matters are usually entertained by local planning authorities. Various national commissions may hold public hearings to address large-scale projects.

In each country, efforts are made to reduce the visual impact of a power station. The main method employed is planting trees and providing artificial earth banks to produce as natural a landscape as possible. Selected colors and textures may be used, either in stone or other materials, to blend structures more readily into the landscape.

It appears that the countries addressed have few statutory or voluntary rules concerning the planning of power station projects. A prevailing feeling seems to be that specific esthetic rules are undesirable and that each case should be dealt with on its merits.

The countries also have not categorized their land surface on the basis of esthetic amenity as such. However, various portions of the countryside that have esthetic value are in use as parks, wilderness areas, etc. It appears that the planning of power plant projects is carried out either to avoid these areas or to minimize amenity damage if siting must take place in such categorized or protected areas.

Protection of visual amenities can cause significant problems to certain individuals. Those living in areas of high visual amenity are likely to be deprived of a supply of electricity if total prohibitions on power plant projects exist in areas of high visual amenity. In the United Kingdom there are individuals living in areas of high amenity who are unable to have a supply of electricity brought in by overhead line and must either contribute to the cost of underground cable, or provide their own small sources of generation.

None of the countries addressed has a clear mechanism to express the value of an esthetic amenity in financial terms. However, it appears that in planning a project in such a way as to prevent harm to an amenity, or at least to reduce damage, money is spent to achieve the least harm.

APPENDIX B

INTERNATIONAL COOPERATION

The countries addressed in this book exchange information on control of pollution from electric power plants among themselves and with other nations, and conduct cooperative studies within the framework of their participation in various international organizations (Table B-1).

In 1971 the Economic Commission for Europe (ECE) Committee on Electric Power established a group of experts on the relationship between electricity and the environment. Since that time, the committee has issued a series of reports prepared for general distribution by the group of experts. To date there are seven documents in the series, Electric Power and the Environment [ECE 1976a,b,1977a,b,c].

The European Economic Community (EEC) has established an environmental policy aimed at preventing and reducing pollution. Toward this end the EEC has established ambient air and water quality criteria and emissions criteria that serve as guidelines for the adoption of standards by member countries. The EEC also serves as a forum for the exchange of technical information and data (e.g., from monitoring networks) and for the adoption of common methodology for measuring parameters and estimating the cost of antipollution measures [Johnson 1974].

The International Energy Agency (IEA) promotes cooperation among member countries to reduce excessive dependence on oil through conservation, development of alternative energy sources, and energy research and development (R&D) and demonstration. With regard to environmental protection, IEA has sponsored research directed toward more efficient combustion and air pollution control technologies for power generation. Its Economic Assessment Service conducts studies of the costs of pollution control equipment and tabulates overall R&D expenditures in member countries. The IEA is initiating an assessment of the economics of environmental controls. The IEA is an autonomous body within the Organization for Economic Cooperation and Development (OECD) [IEA 1981a].

OECD promotes the development of effective environmental policies in member countries. Recently, it has directed its attention to environmental management. Its Environmental Directorate has conducted numerous case studies of energy standards and pollution control methods in member countries.

Table B-1. Participation in International Organizations that Share Environmental Protection Information Relevant to Electric Power [Rubin 1981]

	Organization[a]			
Country	**ECE**	**EEC**	**IEA**	**OECD**
Federal Republic of Germany	X	X	X	X
France	X	X		X
United Kingdom	X	X	X	X
United States	X		X	X

[a] Membership as of January 1980.

APPENDIX C

PROJECTED ENVIRONMENTAL PROTECTION COSTS FOR A UNITED KINGDOM POWER STATION CONSTRUCTED IN 1990

For a modern 2000-MW coal-fired station for construction in 1990, costs (at March 1982 prices) for precipitation and ash handling, would be £30 million, £5 million for a chimney and £30 million for cooling towers (if needed). The total station costs would be £800 million. About £50,000–60,000 would be required for air pollution monitoring instrumentation outside the chimney, and about £50,000 for flue dust monitoring. The operating costs associated with these expenditures are not available. Estimation of noise control costs is very difficult because many noise control properties overlap with others, such as thermal insulation or weather protection, but about £1 million would seem a reasonable total sum, making pollution control costs about 10% of the total capital cost [Barrett 1982].

A recent study of the cost of equipping a 2000-MW station with a Wellman-Lord flue gas desulfurization (FGD) system for 90% SO_2 removal revealed a capital cost (1981 prices for 1990 installation) of £106–122 million, an increase of 15–17% in specific capital costs. The increase in annual electricity generating costs of the system would be £33–54 million, depending on when the station was planned and specific operating conditions. This represents an increase of at least 9% on a station's basic generation costs. It is equivalent to a cost of £320/ton of SO_2 removed [Barrett 1982].

Other methods of SO_2 abatement exist, but have not been costed to the detail of the Wellman-Lord FGD for this system. It is reasonable to assume that for all FGD systems the cost of removing one ton of SO_2 will be around £150–320, for superficial coal cleaning of all coals around £200–278, and for the cleaning of high-sulfur coals around £96–306 [Barrett 1982].

REFERENCES

Barrett, G. (1982) Central Electricity Generating Board, United Kingdom, Personal communication.

Berlin, E., C. Cicchetti and W. Gillen (1974) *Perspective on Power: A Study of the Regulation and Pricing of Electric Power* (Cambridge, MA: Ballinger Publishing Company).

Bich, T. T. N. (1977) "An Econometric Analysis of the Role of Air and Water Residuals in the Production Technology for Steam-Generating Electric Plants," PhD Dissertation, State University of New York at Binghamton.

Brown, R. (1979) "Health and Environmental Effects of Coal Technologies: Background Information on Processes and Pollutants," MTR-79W15901, The MITRE Corporation, McLean, VA.

Bureau of National Affairs (1978) "Federal Republic of Germany," *Int. Environ. Reporter* 241:0101-0105.

Bureau of National Affairs (1980) "United Kingdom of Great Britain and Northern Ireland," *Int. Environ. Reporter* 291:0101-0104.

CEQ (1978) "Environmental Quality 1978: Ninth Annual Report," Council on Environmental Quality, Washington, DC.

CEQ (1979) "Environmental Quality 1979: Tenth Annual Report," Council on Environmental Quality, Washington, DC.

CEQ (1980) "Environmental Quality 1980: Eleventh Annual Report," Council on Environmental Quality, Washington, DC.

Cohn, H. B. (1975) "Environmentalism—Costs and Benefits," *Public Utilities Fortnightly* (July 31), pp. 17–21.

Commission of the European Communities (1979) "State of the Environment. Second Report," European Economic Community, Brussels, Belgium.

Commission on Energy and the Environment (1981) *Coal and the Environment* (London, UK: Her Majesty's Stationery Office).

Congressional Research Service (1981) "Air Quality Management in Selected Countries," Serial No. 97-5, Library of Congress, Washington, DC.

Corps of Engineers (1981) "Final Environmental Impact Statement, Bowline Point Generating Station, Haverstraw, New York," New York District, New York.

Data Resources,Inc. (1981) "The Macroeconomic Impact of Federal Pollution Control Programs: 1981 Assessment," Lexington, MA.

Department of the Environment (1978) *Pollution Control in Great Britain: How It Works*, Pollution Paper No. 9, 2nd ed. (London, UK: Her Majesty's Stationery Office).

Department of the Environment (1979) *The United Kingdom Environment 1979: Progress of Pollution Control*, Pollution Paper No. 16 (London, UK: Her Majesty's Stationery Office).

DOE (1977) "Statistics of Privately Owned Electric Utilities in the United States, 1976, Annual Classes A and B Companies," DOE/EIA-0044(76), Energy Information Administration, U.S. Department of Energy, Washington, DC.

DOE (1978) "Statistics of Privately Owned Electric Utilities in the United States, 1977, Annual Classes A and B Companies," DOE/EIA-0044(77), Energy Information Administration, U.S. Department of Energy, Washington, DC.

DOE (1979) "Statistics of Privately Owned Electric Utilities in the United States, 1978, Annual Classes A and B Companies," DOE/EIA-0044(78), Energy Information Administration, U.S. Department of Energy, Washington, DC.

DOE (1980) "Statistics of Privately Owned Electric Utilities in the United States, 1979, Annual Classes A and B Companies," DOE/EIA-0044(79), Energy Information Administration, U.S. Department of Energy, Washington, DC.

DOE (1981a) "Statistics of Privately Owned Electric Utilities in the United States, 1980, Annual Classes A and B Companies," DOE/EIA-0044(80), Energy Information Administration, U.S. Department of Energy, Washington, DC.

DOE (1981b) "Energy Data Report: Inventory of Power Plants in the United States, 1980 Annual," DOE/EIA-0095(80), Energy Information Administration, U.S. Department of Energy, Washington, DC.

ECE (1976a) *Electric Power and the Environment, Volume III, A Comparative Study of the Legal Obligations Relating to the Environment to be Fulfilled in the Planning, Construction, and Operation of Thermal Power Stations, in the Construction of High Tension Transmission Lines, and Distribution of Electric Power*, ECE/EP/19 (New York: Economic Commission for Europe, United Nations).

ECE (1976b) *Electric Power and the Environment, Volume I, Risk of Shortage in the Supply of Electric Power in Coming Years Resulting from Delays in the Construction of Electric Power Plants, Technical Requirements Related to Environmental Protection, Insufficient Co-ordination with Environmental Authorities at the Planning Stage, and Public Opinion*, ECE/EP/19 (New York: Economic Commission for Europe, United Nations).

ECE (1977a) *Electric Power and the Environment, Vol. V, Desulphurization Policies Affecting the Production of Electric Power*, ECE/EP/19 (New York: Economic Commission for Europe, United Nations).

ECE(1977b) *Electric Power and the Environment, Volume VI, Integration of Electric Power Installations in the Environment*, ECE/EP/19 (New York: Economic Commission for Europe, United Nations).

ECE (1977c) *Electric Power and the Environment, Volume IV, Environmental Problems Caused by the Production of Nuclear Origin*, ECE/EP/19 (New York: Economic Commission for Europe, United Nations).

ECE (1981) *Effects of Thermal Discharge from Electric Power Stations on Water Bodies and Watercourses, and the Standards in Force in Various Countries*, ECE/EP/40 (New York: Economic Commission for Europe, United Nations).

EDF (1982) Electricité de France. Unpublished data.

Elkin, R. (1979) "The Public Must Know—Legislation Is the Biggest Reason for Rate Increases and the Cooperatives Have an Obligation to So Inform Consumers," *Congressional Record* (February 1), pp. E331–E332.

EPA (1978) "Energy/Environment Fact Book," EPA-600/9-77-041, Office of Research and Development, U.S. Environmental Protection Agency, Washington, DC.

EPA (1979) "The Cost of Clean Air and Water, Report to Congress," EPA-230/3-79-001, U.S. Environmental Protection Agency, Washington, DC.

EPRI (1980a) "Potential Impact of R&D Programs on Environmental Control Costs for Coal-Fired Power Plants," Electric Power Research Institute, Palo Alto, CA.

EPRI (1980b) "Research and Development by the Electric Utility Industry," P-1597-SR, Electric Power Research Institute, Palo Alto, CA.

EPRI (1981a) "1980 Annual Report," Electric Power Research Institute, Palo Alto, CA.

EPRI (1981b) "1981–1985 Research and Development Program Plan," Special Report P-1726, Electric Power Research Institute, Palo Alto, CA.

Federal Republic of Germany (1981) "Summary of a Report on Air Pollution by Sulfur Dioxide," paper presented at the Third Seminar on the Desulfurization of Fuels and Combustion Gases, United Nations Economic Commission for Europe, Salzburg, Austria, May 18–20.

Flowers, B. (1976) *Air Pollution Control: An Integrated Approach*, Fifth Report by the Royal Commission on Environmental Pollution (London, UK: Her Majesty's Stationery Office).

GAO (1980a) "Electricity Planning—Today's Improvements Can Alter Tomorrow's Investment Decisions," EMD-80-112, General Accounting Office, Washington, DC.

GAO (1980b) "Construction Work in Progress Issue Needs Improved Regulatory Response for Utilities and Consumers," EMD-80-75, General Accounting Office, Washington, DC.

GAO (1980c) "Electric Power Plant Cancellations and Delays," EMD-81-25 General Accounting Office, Washington, DC.

GAO (1981a) "Electric Power: Contemporary Issues and the Federal Role in Oversight and Regulations," EMD-82-8, General Accounting Office, Washington, DC.

GAO (1981b) "The Effects of Regulation on the Electric Utility Industry," EMD-81-35, General Accounting Office, Washington, DC.

ICF Inc. (1980) "The Economic and Financial Impacts of Environmental Regulations on the Electrical Utility Industry," Washington, DC.

IEA (1981a) "Energy Policies and Programmes of IEA Countries. 1980 Review," International Energy Agency, Organization for Economic Cooperation and Development, Paris, France.

IEA (1981b) "Annual Report on Energy Research, Development and Demonstration: Activities of the IEA 1980/81," International Energy Agency, Organization for Economic Cooperation and Development, Paris, France.

International Institute for Applied Systems Analysis (1981) *Energy in a Finite World: A Global Systems Approach* (Cambridge, MA: Ballinger Publishing Company).

Jansen, P. (1981) "Protection of the Environment—Policy for the Future: Methods and Action in the Federal Republic of Germany," SO 3-81(e), Inter Nationes, Bonn, FRG.

Johnson, S. P. (1974) "Pollution Control in the European Economic Community," in *Energy, Europe and the 1980's*, Conference Publication No. 112 (London, UK: Institution of Electrical Engineers).

King, M., and C. Yang (1981) "Future Economy of Electric Power Generated by Nuclear and Coal-Fired Power Plants," *Energy* 6(3):263–275.

Komanoff, C. (1980) "Pollution Control Improvements in Coal-Fired Electric Generating Plants: What They Accomplish, What They Cost," *J. Air Poll. Control Assoc.* 30(9):1051–1057.

Krohm, G. (1980) "Growth and the Cost of Electric Power," *Public Utilities Fortnightly* (December 18), pp. 32–35.

Lawrence, C. (1982) Electric Power Research Institute, Washington, DC. Personal communication.

Leason, D. B. (1974) "Electricity Generation and Environmental Legislation," in *Energy, Europe and the 1980's*, Conference Publication No. 112 (London, UK: Institution of Electrical Engineers).

Loftness, R. L. (1978) *Energy Handbook* (New York: Van Nostrand-Reinhold Company).

Mangun, W. R. (1979) "A Comparative Evaluation of the Major Pollution Control Programs in the United States and West Germany," *Environ. Management* 3(5):387–401.

McGraw-Hill (1981a) *14th Annual McGraw-Hill Survey of Pollution Control Expenditures* (New York: McGraw-Hill Publishing Co.).

McGraw-Hill (1981b) *Historical Pollution Control Expenditures and Related Data*, (New York: McGraw-Hill Publishing Co.).

McGraw-Hill (1981c) "Utility Cleaning Bill: $101-billion by 1987," *Electrical World* 195(11):43–44.

NAS (1977) "Manpower for Environmental Pollution Control," National Academy of Sciences, Washington, DC.

OECD (1972) "Survey of Pollution Control Cost Estimates Made in Member Countries," Environment Directorate, Organization for Economic Cooperation and Development, Paris, France.

OECD (1973) "Report and Conclusions of the Joint Ad Hoc Group on Air Pollution from Fuel Combustion in Stationary Sources," Environment Directorate, Organization for Economic Cooperation and Development, Paris, France.

OECD (1974) "Economic Implications of Pollution Control: A General Assessment," Organization for Economic Cooperation and Development, Paris, France.

OECD (1975a) "Case History from Germany on Use of Criteria Documents in Setting Standards for the Control of Sulphur Oxides," Environment Directorate, Organization for Economic Cooperation and Development, Paris, France.

OECD (1975b) "Report on the Use of Dose and Effects Data in Setting Air Quality Standards to Provide a Basis for the Control of Sulfur Oxides: Experience in Germany, Japan, and the United States," Environment Directorate, Organization for Economic Cooperation and Development, Paris, France.

OECD (1975c) "Case History from United States on Use of Criteria Documents in Setting Standards for the Control of Sulphur Oxides," Environment Directorate, Organization for Economic Cooperation and Development, Paris, France.

OECD (1976a) "Waste Management in OECD Member Countries," Environment Directorate, Organization for Economic Cooperation and Development, Paris, France.

OECD (1976b) "Water Management in France," Environment Directorate, Organization for Economic Cooperation and Development, Paris, France.

OECD (1977) "Siting of Major Energy Facilities," Environment Directorate, Organization for Economic Cooperation and Development, Paris, France.

OECD (1979a) "The Influence of Technology in Determining Emission and Effluent Standards," Environment Directorate, Organization for Economic Cooperation and Development, Paris, France.

OECD (1979b) "The Siting of Major Energy Facilities," Organization for Economic Cooperation and Development, Paris, France.

OECD (1980) "Siting Procedures for Major Energy Facilities: Some National Cases," Organization for Economic Cooperation and Development, Paris, France.

OECD (1981) "The Costs and Benefits of Sulphur Oxide Control: "A Methodological Study," Organization for Economic Cooperation and Development, Paris, France.

Olds, F. C. (1973) "Capital Cost Calculations for Future Power Plants," *Power Eng.* (January), pp. 61–65.

OTA (1982) "Technological Innovation and Health, Safety, and Environmental Regulations," Office of Technology Assessment, Washington, DC.

Perl, L. J. (1978) "The Impact of Environmental Standards on the Electric Utility Industry," paper presented at the Annual Meeting of the Financial Division of the Edison Electric Institute, New York, NY, May 19.

Ramsay, W. (1979) *Unpaid Costs of Electrical Energy: Health and Environmental Impacts from Coal and Nuclear Power* (Baltimore, MD: The Johns Hopkins University Press).

Rubin, E. S. (1981) "Air Pollution Constraints on Increased Coal Use by Industry: An International Perspective," *J. Air Poll. Control Assoc.* 31(4): 349–360.

Rutledge, G. L., and B. D. O'Connor (1981) "Plant and Equipment Expenditures by Business for Pollution Abatement, 1973–1980, and Planned 1981," *Survey Current Business* (June), pp. 19–25,30,72.

Scharer, B. (1980) "Luftverschmutzung durch Schwefeldioxid," UmweltsBundes-Amt, Berlin, FRG.

Scott, D. (1973) *Pollution in the Electric Power Industry: Its Control and Costs* (Lexington, MA: Lexington Books).

Scott, D. (1976) *Financing the Growth of Electric Utilities* (New York: Prager Publishers).

Service de l'Environnement Industriel (1977) "Installations Registered for Purposes of Environmental Protection Act No. 76-663 of July 19, 1976 and Decree No. 77-1133 of September 21, 1977," Ministere de L'Environnement et du Cadre de Vie, Paris, France.

Service de l'Environnement Industriel (1979) "Industrialization and Environmental Protection in France," Ministere de L'Environnement et du Cadre de Vie, Paris, France.

Simek, H. (1979) "Environment Protection in the Federal Republic of Germany," SO 3-79(e), Inter Nationes, Bonn, FRG.

Statistisches Bundesamt (undated) "Statistisches Jarhbuch fur die Bundesrepublik Deutschland," Wiesbaden, FRG.

Temple, Barker and Sloane, Inc. (1977) "Economic Analysis of Section 316(b) Regulations of the Electric Utility Industry," Wellesley Hills, MA.

UmweltsBundesAmt (1980) English abstract of "Report on Air Pollution by Sulfur Dioxide—Origin, Effects, Reduction," Berlin, FRG.

UmweltsBundesAmt (1982) Personal communication.

Weber, E. (1981) "Air Pollution Control Strategy in the Federal Republic of Germany," *J. Air. Poll. Control Assoc.* 31(1):24–29.

BIBLIOGRAPHY

Alexander, M., and J. Livingston (1973) "What Are the Real Costs and Benefits of Producing 'Clean' Electric Power?" *Public Utilities Fortnightly*. (August 30), pp. 15–19.

Anderson, F., A. Kneese, P. Reed, S. Taylor and R. Stevenson (1977) *Environmental Improvement Through Economic Incentives* (Baltimore, MD: The Johns Hopkins University Press).

Brosch, D. (1976) "Legal Requirements for Approval of Nuclear Power Plants in Federal Republic of Germany," *Energie* 28:284–287.

Burrows, P. (1980). *The Economic Theory of Pollution Control* (Cambridge, MA: The MIT Press).

Cheremisinoff, P. N., and R. A. Young (1975) *Pollution Engineering Practice Handbook* (Ann Arbor, MI: Ann Arbor Science Publishers, Inc.).

Commoner, B., H. Boksenbaum and M. Corr (1975) *Energy and Human Welfare—A Critical Analysis: Volume I. The Social Costs of Power Production.* (New York: MacMillan Publishing Co.).

Congress of the United States (1974) "Hearings Before the Joint Economic Committee on the Economic Impact of Environmental Regulations," U.S. Government Printing Office, Washington, DC.

Council on Environmental Quality (1973) "Energy and the Environment: Electric Power," Washington, DC.

Department of Energy (1980) "Leading Trends in Environmental Regulation That Affect Energy Development," DOE/EV-01682, Office of Technology Impacts, Washington, DC.

Department of Energy (1980) "Governmental Actions Affecting the Environment and Their Effects on Energy Markets," EIA-0201/12, Energy Information Administration, Washington, DC.

Department of Energy (1981) "Environmental-Control Strategies for Coal-Based Industrial Boilers: Analysis of Performance and Costs," prepared by the Argonne National Laboratory, Chicago, IL.

Dibelius, N. R., and W. F. Marx (1973) "Environmental Legislation: How It Affects Fossil-Fuel-Fired Power-Generating Equipment," *Mech. Eng.* (March), pp. 76–80.

Dunkerley, J. (1980) "Trends in Energy Use in Industrial Societies," Research Paper R-19, Resources For the Future, Washington, DC.

Economic Assessment Service (1979) "The Economics of Coal-Based Electricity Generation," EAS Report No. E1/79, International Energy Agency, London, UK.

Economic Assessment Service (1980) "The Economics of Electricity from Coal, Nuclear, and Wind Energy," EAS Report No. H1/78, International Energy Agency. London, UK.

Edel, M. (1973) *Economics and the Environment* (Englewood Cliffs, NJ: Prentice-Hall, Inc.).

Electric Power Research Institute (1977) "Coal Fired Power Plant Capital Cost Estimates," EPRI AF342, Palo Alto, CA.

Engineering News Record (1978) "Largest Costs of Saving Environment Lie Ahead," *Eng. News Record* (June 29), pp. 17–45.

Environmental Protection Agency (1980) "Economic Analysis for the Proposed Revision of Steam-Electric Utility Industry Effluent Limitations Guidelines," EPA 440/1-80/029-e, Office of Planning and Evaluation, Washington, DC.

Frankel, M. (1978) *The Social Audit Pollution Handbook: How to Assess Environmental and Workplace Pollution* (London, UK: The MacMillan Press, Ltd.).

Garvey, G. (1972) *Energy, Ecology, Economy: A Framework for Environmental Policy* (New York: W. W. Norton and Co.).

Golden, J., R. P. Ouelette, S. Saari and P. N. Cheremisinoff (1979) *Environmental Impact Data Book*, (Ann Arbor, MI: Ann Arbor Science Publishers, Inc.).

Gottlieb, M., D. Livengood and K. Wilzbach (1978) "Environmental Control Implications of Increased Coal Utilization under the National Energy Plan," ANL/EES-CP-10, Argonne National Laboratory, Argonne, IL.

Jimeson, R. and R. Spindt (1973) *Pollution Control and Energy Needs* (Washington, DC: American Chemical Society).

Johnson, S. P. (1979) *The Pollution Control Policy of the European Communities* (London, UK: Graham and Trotman, Ltd.).

Joskow, J. (1971) "Meeting the Costs of Environmental Control," *Public Utilities Fortnightly* (November 25), pp. 31–35.

Kneese, A., S. Rolfe and J. Harned (1971) *Managing the Environment: International Economic Cooperation for Pollution Control* (New York: Praeger Publishers).

Kneese, A. and C. Schultze (1975) "Pollution, Prices, and Public Policy," The Brookings Institution, Washington, DC.

Office of Science and Technology (1970) "Electric Power and the Environment," Executive Office of the President, Washington, DC.

Organization for Economic Cooperation and Development (1973) "Analysis of Costs of Pollution Control," Environment Directorate, Paris, France.

Organization for Economic Cooperation and Development (1976). "Pollution Charges: An Assessment," Paris, France.

Organization for Economic Cooperation and Development (1977) "Energy Production and Environment," Director of Information, Paris, France.

Organization for Economic Cooperation and Development (1981) "Energy Research, Development and Demonstration in the IEA Countries," International Energy Agency, Paris, France.

Ponder, W. and R. Stern (1976) "SO_2-Control Methods Compared," *Oil Gas J.* (December 13), pp. 60–68.

de Reeder, P. L. (1978) *Environmental Programmes of Intergovernmental Organizations* (The Hague, The Netherlands: Martinus Nijhoff).

Sharefkin, M. (1974) "The Economic and Environmental Benefits from Improving Electrical Rate Structures," EPA-60015-74-033, Office of Research and Development, U.S. Environmental Protection Agency, Washington, DC.

Thompson, R. G. (1978) *The Cost of Energy and a Clean Environment* (Houston, TX: Gulf Publishing Company).

Walter, I. (1976) *Studies in International Environmental Economics* (New York: John Wiley and Sons, Inc.).

INDEX